BEFORE THE BIG BANG

ALSO BY BRIAN CLEGG

Upgrade Me

The God Effect

A Brief History of Infinity

Light Years

BEFORE
THE BIG BANG

BRIAN CLEGG

ST. MARTIN'S GRIFFIN ❦ NEW YORK

www.stmartins.com

Book design by Gretchen Achilles

The Library of Congress has cataloged the hardcover edition as follows:

Clegg, Brian.
      Before the big bang : the prehistory of our universe / Brian Clegg.—1st ed.
          p. cm.
      Includes bibliographical references.
      ISBN 978-0-312-38547-7
      1. Cosmology. 2. Big bang theory. I. Title.
   QB981.C627 2009
   523.1—dc22

                                                                    2008046035

ISBN 978-0-312-68028-2 (trade paperback)

10 9 8 7 6 5 4 3

DEDICATED TO THE CENTER OF MY UNIVERSE—

GILLIAN, CHELSEA, AND REBECCA

CONTENTS

ACKNOWLEDGMENTS

Thanks to the many people who have provided information including Dr. Marcus Chown, Professor Günter Nimtz, the librarians of Swindon Central Library and the British Library, the Mullard Radio Astronomy Laboratory at the University of Cambridge, and the organizers of the Andrew Chamblin Memorial Lecture.

As always, this book would not have been possible without the help and support of my editor, Michael Homler, and my agent, Peter Cox.

1.

BIG BANG PRIMER

The beginning of things must needs lie in obscurity, beyond the bounds of proof, though within those of conjecture or of analogical inference.

—ASA GRAY (1810–88),
"Darwin on the Origin of Species" (*The Atlantic Monthly*, July 1860)

Welcome to the universe.

"The universe" is an awesome concept, one that belies the apparent simplicity of the term. It's everything, the whole of what's out there, the sum total of existence. We are part of that whole, yet the vast majority of the time we ignore everything but our own tiny corner of it, the infinitesimal speck that is our planet.

Since intelligent reasoning began, human beings have wondered about what the universe is and from where it came. As we see later, all kinds of possibilities were considered, but it wasn't until the twentieth century that our current widely accepted description of the beginnings of the universe—the Big Bang—was first formulated.

The scientific curiosity that makes us wonder about the universe and where it came from seems to be a natural human trait, although it is often suppressed by peer pressure. All children have a sense of wonder when looking at the world around us. They want to know why and how and what, sometimes asking these questions so frequently that adults are driven to distraction. Sadly it's not cool to be interested in science in your teens, so for many that fascination in what's out there gets pushed to one side. But it's still there, waiting to be uncovered.

There's a good reason for this curiosity. As I describe in my book *Upgrade Me*, it was our ancestors' ability to see beyond the present, to ask, What if? to realize that one day they would die, that made them push far beyond their natural evolved capabilities. Our curiosity is part of that ability to look beyond the present moment, and it carries good survival benefits. If we hear a noise at night and ask, "Why did that happen? What caused that noise?" we are more likely to spot a threat before it becomes dangerous. We have an urge to look for causes. We don't, as do many animals, take the attitude that "things just happen"; we know that things have causes and seek them out. This urge to discover causality has some interesting consequences when we ask what came before the Big Bang.

If, as some theories suggest, time started along with space at the Big Bang, then the search for a cause is misleading: in this case there isn't a "before" in which the cause could have taken place. Unless you can take your reasoning outside time and space, as is often done in theological solutions to the question, we have a situation without cause. This is a perfectly possible occurrence when dealing with something as extreme as the origin of the universe, but one with which our causality-seeking brains really can't cope.

The Big Bang is the current, most widely accepted explanation for the origin of the universe, although it must be stressed that it is a best guess, not a proven fact. Inspired by the idea that the universe was expanding (see page 83), the Belgian scientist Georges Lemaître was the first to explicitly mention the idea of the Big Bang (although he didn't call it that). If the universe is getting bigger, as he believed it was, then Lemaître could imagine tracing it back in time, watching it get smaller and smaller until everything was squashed together at the very beginning. This original seed of a universe was originally referred to as a primeval superatom or a cosmic egg.

When I first heard of the Big Bang concept I was dubious about it. In my teens I much preferred the Steady State theory of one of my scientific heroes, Fred Hoyle (see page 116). I was highly disappointed when Steady State was discarded. It felt like a favorite sports team had just lost the championship. It was all very well to imagine everything coming from this compact form, but two problems nagged at me. Why should that initial supercompact universe begin to expand, when all the matter in the universe was pulled together by gravity? And how could you compress everything there was in the universe—that vast quantity of matter—into such a tiny speck?

Initially Lemaître had a poor reception for his ideas. This could be put down to two personal characteristics: he was Belgian and he was a Catholic priest. It was undoubtedly prejudice, but most people thought nothing more impressive than French fries and good chocolate ever came out of Belgium, and Lemaître's membership of the priesthood didn't stand him in good stead in an increasingly atheistic or agnostic scientific community, and one that was particularly suspicious of the Catholic Church's record on the suppression of scientific and cosmological theories. This was, after all, the same church that had squashed Galileo's exploration of the notion that the Earth moved around the Sun. But Lemaître's cosmic egg was disliked for other reasons.

Lemaître had trained at Cambridge University under the great astronomer Arthur Eddington, and although Eddington was very supportive of Lemaître's ideas on the expansion of the universe, he was less happy about the primeval superatom. It seemed to imply that everything originated in a single point, which would involve a massive change in the nature of the universe. This ran counter to his understanding of physics. Others pointed out that Lemaître's picture of the birth of the universe was suspiciously close to the Bible's idea of creation as described in Genesis. Although science should have no problem with religion, rightly or wrongly scientists are always concerned if a theory seems to be inspired by a religious concept.

Fred Hoyle, the champion of the Big Bang's main early rival the Steady State theory, would be the one to give the Big Bang model its name. Until then, what we now call the Big Bang was called the dynamic universe or dynamic evolving model to contrast it with the prevailing idea when the Big Bang first came out of a static unchanging universe. It is generally thought that Hoyle, who first used the name in a popular science radio broadcast on the BBC in 1950, was using the term sarcastically (although he has denied this), but it stuck and has become the accepted name for this dramatic moment of origin of an expanding universe.

In reality, it wouldn't have been too surprising if Hoyle had meant the name as a jibe. After all, if the Big Bang happened, it certainly wasn't big. Lemaître's original version started with a compact superatom containing all matter, whereas the more modern version of the theory has the universe originating from an infinitesimally small point. And it is also often stated that there wouldn't be a bang at all. After all, sound can't travel through empty space. But this complaint about the name is perhaps a little ill-thought through. What space there was at the time was anything but empty; in fact it was packed full of all the matter in the universe, which in principle could have transmitted vibrations corresponding to sound.

That being the case, it's quite possible there was a bang, although of course with no one to hear it, the concept lacks much value. Some cosmologists were against the term "Big Bang" simply because it lacked class (particularly when it was dreamed up by their archrival Hoyle); at the time it was considered unscientific and populist, but to modern ears it is a commonplace. It's a catchy and obvious term. To complain about its triviality when physicists have described particles with properties such as "strangeness" and "charm," and biologists give genes names such as "sonic hedgehog," "grunge," and "INDY" (for "I'm not dead yet," a catchphrase in *Monty Python and the Holy Grail*), seems hypocritical.

Other scientists would fill in more of the picture. How could such a tiny speck result in such a huge universe? Where did the atoms come from of which everything is now made? What was there in the beginning? These are all questions to which increasingly sophisticated answers have been provided. But until recently there is one question that has been purposefully ignored. If there was a Big Bang, what came before?

This is a topic that science has traditionally regarded as taboo, out of reach, and impossible to cover. That might seem a shortsighted view, but one of science's strengths is an awareness of its own limitations. If it is impossible to test a theory against data from experiment or observation, then arguably that theory is not science. This is why many people would argue that science and religion have no great intersection, nor any need to attack each other. Science has no remit to comment on religion, nor should religion attempt to shape science. Religious beliefs are by definition taken on faith without scientific evidence. With no way of proving or disproving a belief, there is no point in trying to take a scientific view of it: it's a null case, as far as science is concerned.

This doesn't mean that science has to dismiss religion, simply that scientific methods can't be used to comment on religious dogma. Similarly, it was argued that, if there were a Big Bang at the birth of our universe, it would be idle speculation to ask what came before. As there was no way to see through the Big Bang into the past of the universe, there was no way of distinguishing among the many theories available, whether a mythical creation story or a piece of science fiction.

However, even at its most sober, cosmology is the most speculative of the sciences, and increasingly testable evidence has become available that is providing some suggestions of what may have come before the Big Bang. No longer is this a subject that is kept firmly on the outside of science, and some of the possible answers to the question are mind-boggling.

Before getting a better understanding of what the universe is and where it came from, it is useful to see how our thinking on the origins of everything have evolved. Looking back through time to an earlier age, there would have been no hesitation in answering the question of what came before the beginning. For many cultures this was obvious: the universe was the work of the creator, so what came before was that creator. However, every culture had its own creation story, each with a different proponent and a different modus operandi. Looking back to the early creation myths can give us a better idea of how humanity came to think about the beginning of everything.

2.

ENTER THE CREATOR

> For a long time it has been known that the first systems of representations with which men have pictured to themselves the world and themselves were of religious origin. There is no religion that is not a cosmology at the same time that it is a speculation upon divine things.
>
> —EMILE DURKHEIM (1858–1917),
> *The Elementary Forms of the Religious Life* (trans. J. W. Swain)

Theories of what came before the beginnings of the universe have been around as long as human beings have been speculating about life, the universe, and everything, an activity that goes back at least 10,000 years and may go much farther.

Broadly, the attempts to explain the origins of the universe fall into three categories: the religious, the philosophical, and the scientific. These first cropped up in that order, but there have been overlaps so that, for instance, there are still widely believed religious explanations in a period when scientific ideas are current.

GOD MADE IT

For many the answer "God made it" is a useful solution to the problem of causality we have if the Big Bang were literally the beginning of time and space. As we have seen, human beings naturally seek causation, so are very uncomfortable with a Big Bang without a cause. However, even children often spot that this isn't really an answer to the philosophical problem. All bringing God into the equation does is shift the causality issue back one level in the hierarchy. They ask, "Yes, but how did God come into being?"

If the answer is, "He was always there," then you have a concept that is no more satisfying in terms of causality than the idea of the universe existing eternally, or springing into being out of nothing with no cause. Note that this doesn't say that either possibility is wrong. I am merely pointing out that "God made it" doesn't solve the causality issue we have, an issue that arises because our brains are wired to assume everything has a cause.

Go back far enough, and the only ideas of where the universe came from are religious. This reflects a variant on Arthur C. Clarke's famous maxim, "Any sufficiently advanced technology is indistinguishable from magic." In this case, it's more, "Anything natural beyond the scope of human conception is indistinguishable from the creation of a god."

We see three broad pictures emerge from the creation myths. Either the universe has always been here and always will be, or it was brought into being out of nothing by a god, or it was brought into being by a god who already inhabited a different universe.

THE GENESIS MYTHS

The most familiar creation myths in the West are those that appear in chapters 1 and 2 of Genesis in the Bible. Before looking at these, I need to be clear about what's meant by myth here, as the word is often used now in a way that suggests it is a derogatory term, and it's not. A myth is a story with a purpose. It tells of something with import for our everyday life, usually occurring far in the past or in a distant land. (George Lucas was deliberately indulging in the language of myth when he set *Star Wars* a long time ago, in a galaxy far, far away.) The myth uses this exotic setting to explain a universal truth, or to put across an important piece of information in a way that will make it easier to remember and absorb.

Although a good many people believe that the Bible is entirely factual, to suggest that the beginning of Genesis is mythical is not inconsistent with the Bible being the word of God. Most Bible scholars see Genesis as a functional myth, and for it to be so fits with the nature of the Bible as a whole. Very little of this collection of books is history. The Bible contains guidance on how to live, love poetry, songs, and religious instruction, but only a few parts correspond even vaguely to the modern idea of history (and even less with an idea of science).

The New Testament, for instance, although superficially historical, contains major inconsistencies between the different gospels. This didn't matter to the writers of these books, because they weren't intending to write history; they were putting across different aspects of the nature of Jesus Christ and his ministry. We know from the gospels that Jesus often used parables—in effect mini-myths—fictional stories with an important point of learning. If the Bible is the word of God as many believe, there is no reason why it too would not contain such illustrative stories, and that is how we ought to see the two descriptions in Genesis of how the world began.

The Jewish people developed their creation myths from earlier Babylonian myths, but with a different intention, changing the form of the stories. The opening of Genesis gives the traditional six-day creation. We begin with the production of the heaven and the earth from nothing; yet to modern eyes confusingly, the Spirit of God moves on the "face of the waters." This reflects a common cosmological idea of the time that everything was created from water, so "the waters" had to have existed before the creation of the universe. Where they came from is not explained.

We then get the introduction of light; the division of the sky from the waters; the division of water and land on the earth; the introduction of plants; the addition of the Sun, Moon, and stars; and finally the introduction of living creatures and humanity. This myth explains God's role as creator, a wholly different function from the subsequent Garden of Eden myth, which is historically incompatible with the first creation story as it puts the creation of man in advance of the animals. This second myth's function is to explain both our role of stewardship on the Earth and the nature of sin.

As it happens, the creation of light before the Sun and the stars, which seemed to early scientists a particularly strange aspect of the Genesis story, does have a reasonable match with current scientific theories. As we see later, it is now thought that the universe was full of light long before any stars were formed. However, to chalk this up as an accurate account in Genesis once more misses the point. The myth's function is not that of science text or literal history.

With this in mind, it becomes much easier to cope with both the diversity and the strangeness of many of the creation myths found around the world. Yes, they can't all be "right" as a scientific description of the beginning, and some of them seem much less likely than the Genesis version to a modern ear, but they were never intended to be an accurate description of real occurrences.

CREATION VARIATION

The Greek myths that would be supplanted in Europe by Christianity began with a state of chaos that has also been described as a void. In one version, within this void was a bird called Nyx that laid a golden egg, out of which came Eros the god of love. The two halves of the shell became Ouranos, the sky, and Gaia, the Earth. These two produced the next generation of gods, the Titans, notably Kronos, who fathered the lesser gods we are familiar with, such as Zeus. In other versions there were more early gods alongside Eros, including Eurynome, the goddess of all things, who consciously drove the emergence of order from chaos.

Straying farther afield, in China a myth tells of the god P'an Ku, also born from an egg, rather handily for the early version of the Big Bang theory that had everything starting from a cosmic egg. The embryonic god was in the egg for many ages; when he was born, the top part of the egg formed the sky and the bottom the ground. P'an Ku then went on, not unlike Slartibartfast in *The Hitchiker's Guide to the Galaxy*, to craft some of the crinkly features of the Earth such as mountains, using a chisel.

Most dramatically of all, P'an Ku himself became much of the rest of creation. Unusually for a god-based myth this happened after his death, aged 18,000, when he provided not only the Sun and Moon with his eyes and the thunder from his voice, but the rivers of the world from his blood. Puzzlingly his skull formed the arc of the sky (which seems to supplant the top of the egg), and his flesh was transformed to soil, quite separately from those chiseled mountains and valleys. We, it seems, were based on his fleas.

This approach puts the P'an Ku myth in a totally different class from the biblical start of the universe. Although both involved a creator god, the Hebrew God was distinct and separate from creation. The Chinese myth makes the god the immediate source of matter, forming the aspects of the universe from his being.

This linking of a god to a physical phenomenon is very common in creation myths, but it's a concept that feels uncomfortable to many modern minds. When I was young I was fascinated by the legends of both ancient Greece and ancient Egypt, yet I found the apparent confusion between clearly animate gods and clearly inanimate aspects of nature confusing. How could anyone see the fiery ball of the Sun as a chariot ridden across the sky by the humanlike Greek sun god Helios? For that matter, how could any god not get intensely bored by a job that involved spending all day, every day, riding a chariot across the sky without an opportunity for a break or a holiday?

Similarly, when I found out about the ancient Egyptian creation myths, I found it bizarre that the sky could be thought of as the goddess Nut, her body forming the curve of the sky above us. After all, this curve didn't even exist (it is a conceptual mathematical form, not a real thing) so how could it possibly be a goddess?

The Egyptian myths are difficult to pin down because their culture did not resort to anything as simple as a god per object or concept. The Sun, for instance, was a number of different gods or parts thereof in its different aspects. But it isn't surprising, given the total dependence of the ancient Egyptian civilization on the Nile, that their creation myths often began with a watery chaos called Nu or Nun. The world as we know it came into being when the waters were divided into upper and lower parts (or when a hill rose out of the waters), on which stood the first of the gods, Atum. He spat out the god of the air, Shu (who would later support Nut), and Tefnut, the goddess of moisture, whose children were Nut, the sky goddess, and Geb, the god of the Earth.

It's the children of Nut and Geb who then became the main heavenly players, Osiris, Isis, Set, and Nephthys, although there would be many more gods along the way. A variant on this myth from the lower kingdom of Egypt ends up similarly, but starts with the sun god Ra emerging from an egg in the ocean and producing those first four gods. Ra would later briefly become the sole creator god in the monotheistic religion of the Aten (the sun's disk) that briefly flourished under Akehnaten.

Excluding Buddhist myths and a few others that parallel them, which posit neither god nor beginning for the universe but rather suggest the whole thing is imponderable (rather a frustrating answer, along the lines of "because" we often use when children ask "Why?"); although mechanisms vary, the general approach is that of a creator or creators, taking the chaotic or formless prebeginning nature of all that was, and turning it into a living, working universe (centered inevitably on the Earth). This emphasis on creation from raw materials is hardly surprising, because experience told those early human civilizations this is what happened.

With most objects we see around us, it's easy to say what came beforehand: the raw materials and whoever shaped them into the object. For a manmade object this is obvious, but it's also true of, say, a mountain range or the Earth itself. The physical forces and matter that went into creating the object came before that object. It's a natural enough assumption, then, that something came before the universe and that this something formed or molded the "wild" nature that existed before into the universe as we know it, just as a human might make bread from wild grain.

SEPARATING CREATOR AND MECHANISM

Although the philosophical revolution largely emerging in the West from ancient Greece enabled humans to look at the universe in a more abstract way, without specifically invoking a creator, the philosophical and religious concepts still remained in parallel. Similarly, even as Newton was putting together an early scientific picture of a mechanical universe, there was no reason to doubt the mythical origins that came from the Bible. As late as the nineteenth century, it was still common to put in the Bible a date for the creation based on calculating the lifetimes of key individual figures in biblical stories, working back to Adam and Eve (see page 65).

However, along the way the picture of what this created universe was had been changing and as those shifts became more extreme, it seemed more and more likely that it would be necessary to move away from a view of the past based on myth, and to try to find a mechanism based on science. This didn't have to remove God from the picture. Many were (and still are) happy to combine a scientific methodology for the beginning of the universe with a divine intervention to make that method happen. All the evidence is, if we are living in God's creation, that He prefers to use good, logical, scientific approaches to the way the smaller things around us work; we have no reason to assume He would take a different approach to the very beginning.

The move to separate religion and an explanation of the nature of the universe in purely knowledge-based (*scientia* in Latin) terms, happened most clearly in the relatively sudden emergence of explanatory philosophy in the ancient Greek civilization around the sixth century BC. This didn't necessarily result in a sudden and complete separation from the picture provided by myth. For example, one version of the ancient Greek mythological cosmology of the time had the universe emerging from a chaos that was a sea of fire.

When Anaximander, an early Greek philosopher from Miletus in Anatolia (now in Turkey) who lived in the first half of the sixth century BC, came up with a "scientific" cosmology (scientific in the sense that he suggested that natural physical forces structured the universe, rather than the gods) he still imagined that universal fire, but made the known universe a sort of shell that protected us from the fire. This shell had holes in it, a big one that formed the Sun and smaller ones for the stars. So the universal light and heat came from the primeval sea of fire in which our island universe floated.

The separation of the mechanisms of nature and the gods was by no means universally accepted in Greece. As in so many other aspects of Greek philisophy, it was Aristotle's view that came to dominate and in this case forced through the correctness of making a separation. Aristotle was born in 384 BC in Stagirus in northern Greece and at the age of seventeen joined Plato's Academy in Athens, where he was to remain for twenty years. The word "academy" has become sufficiently part of the normal vocabulary that it is worth pointing out that this was the original.

Plato, born in 427 BC in Athens, had founded his school when he left military service a couple of years before Aristotle's birth. Like many who have served in the military he was none too impressed with the capabilities of politicians, and set up his school with the specific intention of improving the quality of those in public life. The school, based in Athens, was situated in a grove of trees belonging to a man named Academos, hence the name "Academy." By the time Aristotle arrived, the Academy was well established and respected. In fact it would remain operational until 529 AD, a remarkable 900 years of existence; even the universities of Oxford and Cambridge, the oldest English-speaking universities, won't catch up until the twenty-second century.

Aristotle believed that the only source of light in the universe was the Sun. The stars, he thought, just as the Moon, displayed reflected sunlight. Aristotle was aware that the Moon could be eclipsed when the Earth's shadow fell across it, and that this didn't happen to the stars, but didn't consider this a problem. He thought the stars were never dimmed by the Earth's shadow because that shadow didn't extend beyond the planet Mercury, and so had no effect on these more distant reflective light sources.

It's interesting that the medieval protoscientist Roger Bacon (see page 27), who mostly regarded Aristotle as an unquestionable authority, was prepared to argue on the matter of starlight. It seemed inconceivable to him that starlight should operate in this fashion, and it was an unnecessary complication that the eclipses should somehow avoid the stars, even though sunlight could, according to Aristotle, reach them and shine back from them as with mirrors.

Bacon, however, didn't get it all right. He believed that the Moon also glowed of its own accord, not accepting that it could be as bright as it was purely by reflection of the Sun. He did think that sunlight stimulated the Moon's glow (which is why it could be eclipsed or have dark phases) but that the sunlight merely triggered a much more powerful glow that emerged from the Moon itself.

The universe as far as the writers of the early creation myths were concerned was a very small place. It really only encompassed the Earth and a surrounding sky which as an entity seemed to be a thin skin above the Earth's surface. There might have been a bigger void in which this universe floated, but that was without meaningful size. By the time the ancient Greek philosopher and engineer Archimedes was prepared to calculate the size of the universe to see how much sand it would take to fill it up (see page 23), it had become bigger—around 1,800 million kilometers across—but was still tiny by modern standards.

THE GOLDILOCKS UNIVERSE

Those who still argue today in favor of a divine creator find one particular aspect of the observed universe particularly encouraging toward their belief. A quite remarkable set of circumstances have had to come into play for us to exist at all.

First there are the basic parameters of the universe. Although some of the constants that describe how the universe works could be varied reasonably widely without catastrophic effects (the speed of light, for instance), many are so finely tuned that it would only take tiny changes before life as we know it became impossible. The forces that influence matter, for instance, from the strong nuclear force to gravity, couldn't be varied much before a stable environment became impossible.

Similarly, if neutrons, the neutrally charged particles in the atomic nucleus, were less than one percent lighter, protons, the other components of the nucleus, would decay into a neutron and a positron, making the formation of atoms as we know them impossible. There would be no matter at all. And just tiny variations in the level of something called quantum fluctuations, which we will hear much of later, would mean that the galaxies would never have formed.

Then there are the circumstances of the Earth itself. It sits in a narrow zone where life as we know it is possible. A little closer to the Sun, and Earth would be too hot for life to have emerged: a little farther out, too cold. Even in its current orbit the planet would be too cold for human beings to have developed if it weren't for the greenhouse effect. We're used to the greenhouse effect being mentioned as the bad guy in global warming, and it's certainly true that increases in greenhouse gases are a bad thing, but without the greenhouse effect existing at all, things would be very unpleasant.

The greenhouse effect prevents some of the energy that the Earth receives from the Sun from being reflected back out into space. Gases such as water vapor, carbon dioxide, and methane in the atmosphere allow light from the Sun through, but when the resultant heat tries to escape from the Earth, some of it is absorbed by the gases and reradiated toward the Earth. If there were no greenhouse effect, the Earth would have average temperatures of −18 degrees centigrade, around 33 degrees Celsius colder than it actually is (that's 60 degrees Fahrenheit below current temperatures).

Similarly, the Earth would not have to be much smaller before it was incapable of holding on to enough oxygen for life to develop. If we hadn't got our unusually large Moon, acting as a giant gyroscope to steady us, the Earth would never have had stable enough climate conditions for life to arise. It all seems to add up to an existence that is so carefully balanced that it seems quite natural to ask how this could possibly have happened.

There has been some challenge to this idea of Earth being finely tuned for life. Fred Adams of the University of Michigan in Ann Arbor, for example, has shown with a very simple model of star formation that around a quarter of all possible universes would have energy sources that could support the development of life. However, this doesn't mean that all the other parameters enabling life to exist would be fulfilled in all these possible universes.

We live in such a "Goldilocks" environment—not too hot, not too cold, just right—that it has seemed natural to many that we live in a designed universe, one where our surroundings have been engineered to meet human requirements. This is certainly the understanding of Christian creationists, who believe that the creation story in Genesis is literally true. Deducing an act of creation from the finely tuned conditions that make life possible is called the "strong anthropic principle," which says that the circumstances of our existence are so unlikely otherwise that there has to be some guiding force behind our being here.

Few scientists accept this idea, although they don't object to the weak anthropic principle, which merely says, "We are here, so the circumstances had to be right for us to be here, or we wouldn't be here to observe them." There is inescapable logic to this, although the weak anthropic principle is in effect a circular argument. You can't realistically do anything with it as a scientific theory, as it boils down to, "We have to be able to be here, because we are here."

For the moment it is enough for us to be aware that our drive to discover why and our expectation of causality have historically produced a wide range of myths, some of frightening complexity, to explain where the universe came from. Science also has its explanations, but before we rush across the bridge from cosmological myths to the scientific stories of the very beginning, we need to know just what it is we are dealing with: what the universe is.

3.

WHAT AND HOW BIG?

Don't let me catch anyone talking about the Universe in my department.

—ERNEST RUTHERFORD (1871–1937),
quoted in *Sage: A Life of J. D. Bernal* (Maurice Goldsmith)

What do we mean by "the universe"? How big is this thing? How long has it been around? These are basic questions that are often taken for granted. In many books and Web sites, for instance, you will see that the universe is around 13.7 billion years old, but it's rare that we hear just where that figure came from and the doubts that are attached to it. Without an understanding of what earlier ideas preceded the current best guess, and how the figure of 13.7 billion years was reached, we can't deal with the science of what came before.

WHAT IS A UNIVERSE?

Of those three big questions in the first paragraph, the one that gets most often overlooked is, "What do we mean by the universe?" surely an essential starting point. According to my dictionary, the word "universe" comes from the Latin for "one turn" and refers to everything existing taken together, all things considered as a systematic whole. It's everything physical, from space itself to every star, planet, creature, atom, and photon of light out there.

Although that definition has been pretty well consistent through time, the vision of what that universe encompassed has passed through four broad phases. In prephilosophical times it was generally considered to cover "the earth and the heavens" one way or another, a combination of the clear, practical division of down here on Earth, and the untouchable stuff out there that surrounds us, whatever it may be.

From the philosophical approach of the ancient Greeks onward, that picture became more subtle and corresponded roughly to what we now regard as the solar system. Although it was still "down here" and "up there," two unmixable arenas that were divided by the Moon into the imperfect worldly sublunar and the perfect and ethereal superlunar or quintessence, there was a clear structure to "up there" rather than just "the heavens." And although that structure had the Earth at its center rather than the Sun, it was still a crude version of the solar system, finished off with the stars in a sphere beyond the orbits of Jupiter and Saturn.

As with much of ancient Greek science, this idea of the universe held all the way through to the Renaissance and it wasn't really until the mid-eighteenth century that a bigger picture emerged, an idea of the universe that was more like our current understanding of our galaxy, the Milky Way. And finally, in the twentieth century it was realized that the fuzzy nebulae that could be seen among the stars were not gas clouds but galaxies in their own right, transforming our understanding of the nature and scale of the universe once more.

FILLING THE UNIVERSE WITH SAND

As understanding of the nature of the universe changed, so did ideas of its size. There were various attempts in ancient Greek times to come up with a measurement of the universe, but none was more interesting than that of Archimedes, particularly so because this outstanding mathematician and engineer was not looking for answers to the great questions of life, the universe, and everything, but instead engaged on what seems a totally trivial exercise: working out how many grains of sand it would take to fill the whole of space.

We know little that is factual about Archimedes' life, but he is thought to have died in the Roman attack on Syracuse, and if, as legend has it, he was around seventy-five at the time, this would put his birth date around 287 BC. At the time of his death, Archimedes was best known for his feats of engineering. For him, according to the Greek writer Plutarch (writing a good 350 years later), these were "diversions of geometry at play," but the defenders of Syracuse would have seen things very differently. They were said to have used a range of mechanical engines devised by Archimedes to attack Roman ships, and even to have contemplated (but never built) a sun-powered death ray in the form of huge mirrors to concentrate the sun's light on distant Roman ships and set them alight.

But it was not Archimedes' undoubted mechanical genius, nor his scientific work that brings him into the story of the scale of the universe, but rather one of the strangest books in history, *The Sand Reckoner*. This is addressed to the king of Syracuse, Gelon. *The Sand Reckoner* opens with a challenge to the king to be more imaginative than many of his contemporaries:

There are some, king Gelon, who think that the number of the sand is infinite in multitude . . . but I will try to show you by means of geometrical proofs, which you will be able to follow, that, of the numbers named by me and given in the work which I sent to Zeuxippus, some exceed not only the number of the mass of sand equal in magnitude to the Earth filled up in the way described, but also to that of a mass equal in magnitude to the universe.

Coming up with the number of grains of sand equivalent to the universe wasn't a practical proposition, but this wasn't really Archimedes' intention. The Greek number system was extremely clumsy. The biggest number available was a myriad—10,000—and in this book, Archimedes was setting out to prove that you can extend this system as far as you like, even to deal with such huge numbers as the grains of sand required to fill the universe.

The "universe" he would fill with sand was more like our idea of the solar system—even so, quite a size. Tantalizingly, Archimedes goes a little further. He imagines an even bigger universe, suggested by a theory around at the time that instead of the Sun moving around the Earth, as it appears to do, the Earth moved around the Sun. It's tantalizing because Archimedes' passing mention of this theory of Aristarchus is the only surviving reference to the earliest known person to spot that we move around the Sun rather than the other way around. Archimedes says:

Now you are aware that "universe" is the name given by most astronomers to the sphere whose center is the center of the Earth and whose radius is equal to the straight line between the center of the Sun and the center of the Earth.

This is the common account as you have heard from as-

tronomers. But Aristarchus of Samos brought out a book consisting of some hypotheses, in which the premise leads to the result
that the universe is many times greater than that now called. His
hypotheses are that the fixed stars and the Sun remain unmoved,
that the earth revolves around the Sun in the circumference of a
circle, the Sun lying in the middle of the orbit . . .

Archimedes goes on to do calculations for the size of both the
accepted universe and Aristarchus' quaint alternative. After making a few assumptions such as "The diameter of the earth is greater
than the moon, and the diameter of the sun is greater than the
earth," and some elegant geometry, Archimedes comes to the conclusion that the universe is no more than 10 billion stades across.
That's a measurement based on a stadium, just as we often estimate
in football fields, which was around 180 meters, so it makes his universe 1,800 million kilometers. This is just a little larger than the orbit of Saturn, and thus a remarkably good attempt at coming up
with a size for the universe as it was understood at the time.

IGNORING THE PARALLAX

Strangely enough, Greek philosophers were aware of a piece of evidence that suggests the universe is much bigger than they supposed,
but they chose to use this evidence to support a totally different
viewpoint. What now appears to be a fairly obvious slip-up is something we should all bear in mind when looking at the cosmological theories that hold sway today. It is entirely possible that some
of the indirect evidence behind the generally held models of the
universe could be interpreted entirely differently if we didn't hold
certain assumptions.

The clearest of these assumptions that could disrupt present-day theories is that the basic constants of the universe (the speed of light and the strength of gravity, for example) do not vary in different parts of the universe, nor have they varied through time. If these constants weren't as fixed as we assume (and it is largely an assumption of convenience, although there is some evidence) then it is entirely possible that we could explain many of the apparent properties of the universe without resorting to the strange concepts we meet later in this book such as inflation, black holes, and dark energy.

The piece of evidence that the ancient Greeks knew but misinterpreted, suggesting that the universe was much larger than they thought, was the lack of parallax in the stars. Parallax is the mechanism we use to tell how far away something is, relying on having two eyes, separated by a distance. If you hold up your finger in front of your eyes and flip between closing your left and right eye, you will see the finger move from side to side compared to objects on the other side of the room. This simple geometrical effect of apparently different positioning is part of how our brains decide whether something appears bigger because it is closer to us, or bigger because it really is that way. It's part of our 3D vision mechanism.

Now think of the Earth, taking its annual trip around the Sun. Imagine looking at the skies from the Earth on two occasions six months apart, comparing the view taken on either side of its orbit. The result is like looking at the universe through a pair of eyes that are separated by many millions of miles. Stars that are closer to us should move more as a result of the different viewpoints than stars that are farther away. But we don't see any such relative movement when comparing viewings that are taken six months apart.

However, rather than deduce an immense universe (or more precisely an immense galaxy, as the stars the Greeks were comparing are all in our galaxy), instead, the Greeks assumed this lack of motion helped prove that the Sun moved around the Earth, rather than the other way around. If the Earth were fixed, then it would generate no parallax. So even though one philosopher had suggested the Sun was the center of things, the Greeks rejected his viewpoint.

A DIFFERENT SCIENCE

Parallax wasn't the only reason for the Greek argument against the Earth moving. They also couldn't understand why we didn't feel the wind caused by the planet's motion. And a moving Earth presented a real problem for their ideas on the nature of gravity and motion. Greeks believed that an item only moved if something was pushing it, as opposed to Newton's laws of motion, which say a body stays in motion, or stopped, unless something pushes it. The pull of gravity, they reasoned, was toward the center of the universe. And if that was the Sun, why didn't everything fly off the Earth toward the Sun?

Thanks to this supporting evidence it seemed reasonable to them to believe that the stars were still as originally conceived, set in a sphere on the edge of the known solar system. However, this was by no means the only classical viewpoint. When examining classical cosmology, we have to bear in mind that at the time there was a totally different approach to science (or, rather, natural philosophy). Theories weren't tested against experiments, or observations of the natural world; they were tested by argument.

Two or more philosophers would put up alternative viewpoints on (say) the nature of matter or the extent of the universe. Their arguments would be discussed and pulled apart. Then, whichever argument seemed most reasonable would be considered the accepted view. If this sounds a bizarre means of establishing scientific truth, remember that this is the same process by which we still discover the "truth" in a court of law. Such was the dependence on argument that Aristotle's view that heavier weights fell faster than lighter ones was accepted without bothering to put it to the test for nearly 2,000 years, all the way up to Galileo's time.

Some classical philosophers believed that the universe was infinite in extent. Their arguments remind us of all that is bad about philosophy when we attempt to use philosophical reasoning to deduce scientific facts. The universe, they argued, by definition encompasses everything. That's implicit in the name. Now if it's finite, then it has boundaries. And that has to be the edge between the universe and everything else (whatever that might be). But if there was an "everything else" the universe wasn't everything. This is self-contradictory, so the universe can't have boundaries. And as far as they were concerned, that meant it was infinite. (In practice, as I show later, it is possible for something to be finite but not have boundaries, but you can't blame them for getting it wrong.)

This debate raged on for many years, and it's interesting that unlike the myth-based cosmologies we saw in the previous chapter, it was fairly unusual to bring theological matters into the dispute. Even as profound a Christian as the Franciscan friar Roger Bacon, who began (but never finished) the work of assembling a great encyclopedia of science in the 1260s, resorted to logic rather than theological pronouncements in deciding whether the universe was infinite.

A REMARKABLE FRIAR

For a man of his time, Bacon was well traveled. From his birth in Ilchester, Somerset, in the early thirteenth century he moved to the alien environment of Paris, first learning and then teaching at the university. From there he moved back to England, to Oxford, where he was exposed to natural philosophy and joined the Friars Minor. Becoming a Franciscan proved problematic: Bacon was determined to write about science, but writing books was forbidden for the friars.

He was to get around this by enlisting the support of an important man, Cardinal Guy de Foulques, but even de Foulques had not enough authority to give Bacon permission to write. Bacon was almost resigned to never getting any further when news arrived from Rome. The pope had died and a newcomer, Clement IV, had ascended to the throne of St. Peter. The original name of that new pope was Guy de Foulques. With his sponsor's blessing, Bacon was able to proceed.

Bacon first intended to write a brief outline of an encyclopedia of knowledge, but he found it impossible to hold back the ideas. His so-called letter reached a total of half a million words, more than five times the length of this book. Resignedly, Bacon sent it off to be copied and started on a covering letter for what had now become a major manuscript, but again he was carried away. By the time he was finished he had completed three volumes. Among them they covered philosophy, astrology, astronomy, geography, optics, mathematics, and more. It was a remarkable one-man study, and yet it was, as far as Bacon was concerned, only the beginning.

As he waited for a response to his proposals, he heard news that pitched him from joy into despair. Before his work had reached Rome, Clement IV had died. Under a new strict regime, Bacon seems to have been imprisoned for "suspected novelties." The duration of the imprisonment is uncertain. Estimates range from two to thirteen years before he was returned to Oxford. His books had all been suppressed by the Church and remained so for many years. He died in 1294 at the age of eighty. But his writings would never be forgotten.

ROGER BACON'S UNIVERSE

Bacon's view on the universe comes out of his great book proposal, the *Opus Majus*. Bacon did not argue with the accepted structure of the solar system that put the Earth at its center (it would take Copernicus to do that hundreds of years later), but thought it was much larger than Archimedes allowed. He also considered the shape of the universe. Archimedes had proclaimed that the universe was spherical but presented no argument for this; it was probably just because this seemed the perfect shape. But Bacon wasn't content with an assumption. Thinking through the possible forms the universe might have, he pointed out in the *Opus Majus* that his ancient Greek hero, Aristotle, gave some other possibilities:

> Other figures especially suitable would be either of oval shape or those like it, or of lenticular shape and those formed like it, according to Aristotle in his book On the Heavens and the World. But he states that the heavens do not have a shape of this kind, yet he does not give a reason.

Bacon goes on to explain what he means by "lenticular." These days we would say it meant that an object was lens-shaped, but Bacon goes back to the origins of both "lenticular" and "lens" and comments that "A lenticular shape is that of the vegetable called lentil."

Bacon reckoned that the universe could not be, for instance, "of some angular form" because when it rotated it would leave behind a vacuum and, "Nature does not endure a vacuum." He then went on to dismiss other shapes such as the lens or "the form of a cheese" (presumably a short cylinder) which are symmetrical about one axis, but not about another. He allows that they could rotate safely around that one axis, but there would be problems otherwise. Imagine a circular lens, held by two fingers at the lens's edge, on opposite sides of the center. Start to rotate it around the axis between those fingers. Watch the edge of the lens: wherever it was a moment ago is now empty. If this were the universe, you would have just created a vacuum where the universe used to be.

So, argues Bacon, unless the universe is a sphere there is always the possibility of its rotation creating a vacuum, an uncomfortable nothingness. Although a stationary universe, or a lens or cheese shape rotating around the short diameter wouldn't create a vacuum, to Bacon's peculiarly medieval logic, this wasn't good enough. He believed that nature would "not endure a vacuum nor a possibility of one forming." Such was the impossibility of a vacuum, that there could not be even the chance of it occurring—who was to say that the universe wouldn't move a little on the wrong axis—and so it had to be spherical, the only shape that could rotate in any direction around its center and not leave a vacuum. He, perhaps luckily, did not point out that any shape would cause a vacuum if it moved laterally. The only way of avoiding this would be if the universe were without any shape at all.

BACON'S FINITE VISION

With the universe established to his satisfaction as spherical, Bacon went on to consider whether there could be more than one universe and whether the universe was infinite. Bacon dismissed the existence of multiple universes using the same argument he used to rule out a nonspherical universe. Two universes, a pair of spheres, together form a nonspherical shape. Once more, Bacon argues, rotation will leave behind an unacceptable vacuum. (He doesn't deal with the possibility that one universe totally encloses the other, but arguably this is just one bigger universe.)

As for the size of the universe, Bacon believed that it could not be infinite. He brought in geometry to justify this view, imagining drawing lines from the center of the universe out to its edge.

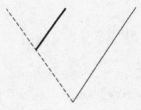

He starts with two lines forming a V, starting from the same point at the center of the universe and stretching to the edge of the immense sphere. Because the universe is spherical, these lines are each of the same length. He then adds a third line (the heavy one in the diagram) starting part way up the left side of the V and parallel to its right-hand side. This also heads off to the edge of the universe. Now, if the universe is infinite, he argued, the two parallel lines should be the same length (presumably because two parallel lines meet at infinity, the strict mathematical way of saying that they never meet). The two sides of the V are of identical length, but the heavy line is also the same length as the part of the broken line from the edge of the universe to where the heavy line meets it, another pair of lines going from the same point to infinity. So, argues Bacon, part of the broken line equals the whole of the broken line. This is an impossibility. So the edge of the universe isn't at an infinite distance at all (and, therefore, the parallel lines weren't the same length after all).

Unfortunately, Bacon had no chance of getting this right, thanks to one of the more counterintuitive aspects of the mathematics of infinity: infinity plus a bit more is infinity, something that he didn't realize. So his attempts at a geometric proof offer no effective arguments against an infinite universe. The reality was to be much more complex and would take many hundreds of years to untangle.

A COLLAPSING UNIVERSE

Whether the universe was finite or infinite, any attempt to think logically about it presented some puzzles for early cosmologists. One of the best-known problems is Bentley's paradox, named after one of Newton's contemporaries, Richard Bentley. Bentley was a scholar, like many of his time, in holy orders, and later Master of Trinity College, Cambridge. Although a classicist and theologian, Bentley spent some time in the 1690s lecturing on Newton's physics, and corresponded with Newton on a number of matters, including his paradox.

Once Newton's ideas of gravitation became widely known, the consequences for cosmology seemed distressing. If all bodies attracted each other as Newton said, Bentley thought that each star and planet should, over time, be pulled toward its neighbors. Eventually the whole universe would come crashing in upon itself. Things would be even worse if the universe were infinitely large. Then the sum of the gravitational force could itself be infinite, making it impossible for bodies such as the Earth and the Sun to exist without being ripped apart in the terrible tide of gravitational forces.

This infinite force argument was less certain than the problem of a collapsing universe, not only because no one knew if the universe was infinite, but also because of a mathematical trick that had been around since ancient Greek times when Zeno described a paradox involving a race between the hero Achilles and a tortoise.

Achilles, arguably the fastest man of his day, the equivalent of a modern sports star, takes on the ponderously slow tortoise in a race. Considering the result of a rather similar race in one of Aesop's fables (roughly contemporary with Zeno's paradoxes), it's not too surprising that the tortoise wins. But unlike the outcome of the race between the tortoise and the hare, this improbable result is not brought on by laziness and presumption. Instead it is the sheer mechanics of motion that Zeno uses to give the tortoise the winner's laurels.

Zeno assumes that Achilles is kind enough to give the tortoise an initial advantage; after all this is hardly a race of equals, and Achilles was a hero. He allows the tortoise to begin some considerable distance in front of him. In a frighteningly small time (Achilles is quite a runner), our track star has reached the tortoise's starting point. By now, though, however slow the tortoise ambles, it has moved on a little way. It still has a lead. In an even smaller amount of time, Achilles reaches the tortoise's new position, yet that extra time has given the tortoise the opportunity to move on. And so the endless race carries on with the logical equivalent of a mathematician's three dots, Achilles eternally chasing the tortoise but never quite catching it.

What this paradox reflects is that it is possible to have an infinite series that adds up to a finite value. Just think of adding up this series:

$$1 + \frac{1}{2} + \frac{1}{4} + \frac{1}{8} + \frac{1}{16} + \frac{1}{32} + \frac{1}{64} \ldots.$$

Each new item in the series takes the total closer to 2, but it never quite makes it. The total of the series at each stage is:

$$1, 1\tfrac{1}{2}, 1\tfrac{3}{4}, 1\tfrac{7}{8}, 1\tfrac{15}{16}, 1\tfrac{31}{32}, 1\tfrac{63}{64}$$

and so on. Imagine you went on for billions and billions of terms to reach the term called n. Then the total would be $1 + ((n-1)/n)$. If a series gets smaller fast enough, then the total will only be a finite number. You can add as many items as you like in this series— an infinite set of items, even—and you will still never get past a total of 2.

Such a limited series could be an excuse for the idea that an infinite universe would not have to exert infinite forces, because of the way gravitational pull reduces as you get farther away from an object. It decreases quickly enough to produce this sort of series, so it may be that an infinite set of stars and planets would still exert only a finite gravitational pull. But that wouldn't solve Bentley's original problem of why everything isn't crashing together in a stellar and planetary car crash of universal proportions.

Newton, in fact, preferred the idea of an infinite universe, and suggested to Bentley that his paradox could be avoided if everything were evenly spread through the universe, so that the pulls in all directions canceled out, leaving everything stable. However, Newton was aware how fragile a picture this was: it only required one planet or star to be shifted a little out of place and the whole thing would crash together with increasing momentum.

For a solution, Newton turned to God. As were most scientists of his day, he was devoutly religious, even if his version of Christianity was anything but orthodox, and he believed this potential instability was not a problem, as God's hand was ever on the universal tiller, able to make constant minute corrections to keep the mechanisms of the universe ticking away. It's a mixed metaphor, but you get the idea.

THE SURPRISING BLACK SKY

Other scientists of the day were beginning to think there was an even worse problem presented by an infinite universe that was uniform in structure: Where are all the stars? When we look out on a dark night without the light pollution that comes from city streets, the main impression is blackness. A sweep of black velvet above us, pierced by a few thousand brilliant points of light. The vast majority of the night sky is black. Yet if the universe went on forever, and stars were evenly distributed throughout it, you should see a star in whatever direction you look. Instead of being primarily black, the night sky should be a uniform star glow, as the light from those infinite sets of stars crashes in on us.

It's hard to say who first identified this problem, as unlike Bentley's paradox it didn't require Newton's ideas on gravity to make it an issue. It seems to have been referred to by both Kepler and Halley, but is now called Olbers' paradox, after an observation by the nineteenth-century German astronomer Heinrich Wilhelm Olbers that was published by his colleague Johann Bode. Born in Arbergen in 1758, Olbers was a medical doctor, but it was his hobby of astronomy for which he is remembered.

Like a number of other astronomical paradoxes (such as why stars twinkle and why the Moon appears to be much larger than it really is, particularly when it's near the horizon), Olbers' puzzle was to continue to cause confusion long after it was officially solved. As late as 1987, a study showed that 70 percent of textbooks gave an incorrect explanation for why the night sky is mostly dark.

This isn't too surprising as it is easy to generate a number of plausible explanations for this effect.

Olbers' own explanation was dust in large quantities. He argued that space is full of dust, and that before you get too far away from the Earth, that dust drastically reduces the intensity of light, so most starlight (and for that matter the heat from all those stars) never makes it to the Earth. This is not a bad argument, although it has since been pointed out that the dust itself would heat up and would eventually glow of its own accord.

Others have argued that red shift is at work. This is something we will cover in more detail later in the book. Red shift is an effect like the Doppler effect, the one that makes a sound such as a police siren change in pitch when its source is moving. We now believe that the universe is expanding, and the farther away you get from us, the faster all the stars are going with respect to us. When a light source moves away from us, it is red-shifted: the light moves farther down the electromagnetic spectrum, having lower energy. With the sort of movement we see in distant galaxies, much of the light will be shifted out of the visible range and thus these far-reaching light sources will not disrupt the blackness.

This red-shift effect is certainly real, but it is not the main reason we don't get an Olbers effect. We now know that light has a finite speed, and assuming the universe has a particular age (see page 64), we can see only light that has been traveling for less time than the age of the universe. So even in an infinite universe, if there is a specific creation time, there will come a point where looking out into space you see nothing, because the light hasn't managed to get from there to here yet. This, now the widely accepted solution, was first proposed by the writer Edgar Allan Poe in "Eureka: A Prose Poem," an essay on "the material and spiritual universe" that he adapted from a lecture on cosmography he gave in New York in 1848.

Ironically, Poe used the explanation not to disprove Olbers' paradox, but to suggest that the universe isn't infinite. He argues that if the stars went on forever we would have uniform illumination. As we don't, he says that the only reason that might be the case if the universe were very large is if much of the light hasn't had time to reach us yet. "This may be so," Poe says, "who shall venture to deny? I maintain, simply, that we have not even the shadow of a reason for believing that it is so."

Although many over the years had speculated that the universe was infinite, for large periods of time it was considered to have a finite compass. It's not surprising there has been confusion, because it's not as if we can take a tape measure out in space and work out accurate dimensions. We need some way to find the size of the universe without getting out there.

MEASURING THE IMMEASURABLE

Look up at the night sky. On a clear dark night, you will see many, many stars. How far away are they? It's impossible to say simply by looking at that array of glittering points. They could be immense and distant, or little bigger than you actually see them, but close up. Back in ancient Greek times, and all the way through to the Renaissance, it was assumed that all the stars were situated somewhere around the edge of the solar system. Archimedes, remember, in his book *The Sand Reckoner* (see page 22) calculated the universe was around 1,800 million kilometers across, which would put the stars just outside the orbit of Saturn.

Over time, however, not only was the scale of the universe found to be much greater than Archimedes suggested, but also it was discovered that the stars can vary hugely in distance from us. Although we see all the stars in a familiar constellation such as the Big Dipper as clustered together, they are vastly different distances away from the Earth. It gets worse when you take into account the bright lights of other galaxies and the stars they contain, where the distances away become literally astronomical. The real problem with a star is that we have no way of directly distinguishing between something that's bright and far away and something that's dimmer and closer. And the farther out into the universe you go, the harder this all becomes.

The first person who really pulled the universe from the Greek "solar system with a sphere of stars" model to something approximating the Milky Way galaxy was William Herschel.

THE MUSICAL ASTRONOMER

Herschel had an unusual background for a scientist. Born in Hanover in 1738, Friedrich Wilhelm Herschel was the son of the bandmaster to the Hanoverian Guard. Not surprisingly, he developed an early interest in music. Herschel joined his father's band at the age of fourteen, but military bandsmen of the time could not expect to stay comfortably in the barracks for too long. Four years later he was sent over to England as part of a national defense force in case of a French invasion (George II was king of both England and Hanover).

When Herschel returned to Hanover, he soon applied for a discharge, which was duly granted. For some reason Herschel has been described as deserting, but there is no evidence to suggest this. Once untangled from the military regime he was eager to get back to music. His time in England had been pleasant; he had picked up a good smattering of English there, and so he joined his brother Jacob in a trip over to London in 1757. It wasn't intended to be a permanent move, but Herschel could hardly have predicted what was to happen to him.

Before long, Herschel's skills at the organ won him a position playing at the Octagon Chapel in Bath, a city that was then at the height of its popularity with the fashionable set. Herschel's talent and cosmopolitan social skills won him the job. When he wasn't playing, he took on private pupils and composed. As a successful musician in the most exclusive resort in the country, Herschel was not short of money, and increasingly had spare time in which to amuse himself. He took up astronomy.

For many wealthy people of the period, astronomy was a casual interest, and this seems to have been the case originally with Herschel when he bought a small telescope and made occasional attempts to scan the heavens, but his real interest was sparked by the thought of making his own telescope. Herschel had no experience of instrument making, but he had enthusiastic helpers in his sister Caroline (who was by now his caretaker) and his brother Alexander. The enthusiasm was needed. It was very easy to fail when every aspect of production, including polishing and shaping a metal plate to a perfect mirror surface, was a matter of trial and error. But by 1774 he had constructed his own telescope, a five-foot-long tube with an eight-inch-wide mirror at the end.

With hours of patient sweeping of the skies, Herschel discovered what he thought was a new comet, but in fact proved to be an undiscovered planet, the first new planet to be found since ancient times, now called Uranus. By now astronomy was an obsession. The king, George III, was sufficiently impressed by Herschel's work to devise a special post for him, King's Astronomer. Herschel could give up music and dedicate himself full time to the skies. But there was a price to pay. He was too far from the court in his comfortable house in Bath and would have to move to a more convenient location. He settled on Slough, near Windsor, where he built his greatest telescope ever, a huge monster with a forty-nine-inch mirror and a tube forty feet long that was mounted in a great wooden structure of poles and ladders, allowing it to be tilted and turned to take in any point in the heavens.

BRIGHTNESS AS DISTANCE

Although Herschel is best known for his discovery of Uranus, his importance here is his later attempts to measure the scale of the universe. To do this, he made a totally unfounded assumption, one that he knew was almost certain to be wrong (and it is), but that made it practical to undertake this exercise. He assumed all stars had the same brightness, so any variation in their magnitude could be put down to differing distances. Although this is a gross over-simplification, and makes any values only accurate to a couple of orders of magnitude, it at least made it possible to get a guesswork picture of the universe.

As many had before him, Herschel had noticed that there were many more stars in a band through space called the Milky Way, which we now know to be our home galaxy. This strip of fuzzy light is much more highly populated than the view if you look straight up or down, at ninety degrees to this band. Herschel suspected from this that we were part of a disk that was the universe.

For his measurements of brightness, he started with Sirius, the dog star, already known to be the brightest star in the Northern hemisphere. Combining this use of Sirius as a guide with the knowledge that light intensity dropped off with the inverse square of distance (so with two identical stars, one three times as far away as the other, the brightness of the more distant star will be $\frac{1}{3}^2 = \frac{1}{9}$ of the closer star), he was able to work up a picture of the scale of the universe based on the distance to Sirius.

This distance itself was unknown, but that didn't faze him. He defined the distance to Sirius as his unit (the siriometer) and measured everything else by comparison with it. From his measurements, he reckoned the universe was around 1,000 siriometers across and 100 siriometers thick. Sirius, also known as Alpha Canis Majoris, is around 8 light-years away, so Herschel's universe was 8,000 light-years in diameter. (A light-year, one of the two standard distance units in astronomy, is the distance light travels in one year, just under 9.5 trillion kilometers.)

Considering the crudeness of Herschel's measurements, the size he calculated doesn't compare too badly with current estimates on the size of the Milky Way galaxy at around 100,000 light-years across. Herschel's observations, although not hugely accurate, gave a feeling for the relative scale of what was then seen as the universe, but didn't tell us how big things were. A refinement of his idea of relative distances would come into use eventually (and is still used today), but there remained the problem of how to fix that initial starting distance. How big was a siriometer? It would take that mechanism that the Greeks had been aware of, but had ignored in thinking about the size of the universe: parallax.

RETURN TO PARALLAX

As we saw earlier (page 25) parallax involves taking a look at a distant object from two different viewpoints. The farther away the object you look at, the less it moves when you change your position. Astronomers can use a large-scale version of looking with one eye and then the other by looking at something in the sky from opposite sides of the Earth's orbit around the Sun. It's not exactly instant— you have to wait six months for the Earth to move into position— but the two observations will be taken around 300 million kilometers apart—not a bad size for a scientific instrument, and providing much more leverage than the ten-centimeter separation of the eyes.

The parallax approach would confusingly result in astronomers adding a second unit of distance to their kitbag, the parsec. This is a contraction of "parallax arcsecond." It is the distance that will result in a parallax shift across half the Earth's orbit of one arcsecond, $\frac{1}{3600}$th of a degree. The equivalent distance is around 3.25 light-years.

Although convenient for astronomers using parallax, the parsec is an unnecessary extra unit. In almost all other areas of science, there is one standard measure that is used in all cases (the standard measure of distance is the meter). Astronomers really haven't caught up with the rest of the scientific world in using these two measures (to make matters worse, they also use the "astronomical unit" or AU which at around 150 million kilometers is an approximation of the distance from the Earth to the Sun), neither of which meets international standards. They tend to use parsecs among astronomers and light-years for communication to the public, but arguably should get their act together!

MOVING THE STARS

The first practical attempt at using parallax on the stars was undertaken by Friedrich Wilhelm Bessel, a contemporary of Herschel's equally talented son John, and a notable German astronomer of the period. Mathematicians and physicists know Bessel best as the man who gave us Bessel functions, equations that are used to understand the behavior of waves and conduction of heat, but for astronomers, Bessel is the man who put a yardstick against the stars.

Born in the Westphalian town of Mindel in 1784, Bessel left school at the age of fourteen and started his career in accountancy, but in his spare time he became more and more fascinated by astronomy and mathematics with more interesting applications than a column of figures in an account book. At the age of twenty-two, after significant coaching from the paradox man, Heinrich Olbers, Bessel took up a post of assistant at the private Lilienthal Observatory near Bremen. Bessel's first distance measurement was of the star 61 Cygni. Using measurements of a tiny shift in position across the six-month shift from one side of the Earth's orbit to the other he was able to say that this star was 100 trillion kilometers away, or 10.5 light-years (we now know it's more like 11.4 light-years; this was very accurate for a first attempt).

With the relatively close stars, distances measured by parallax can be very precise. Unfortunately, as things get farther away, the shift due to parallax gets smaller and smaller. Before long—well before we reach the distance of another galaxy—it just doesn't work anymore. Now we get to one of the big guesses used by astronomers: standard candles. If this term seems surprisingly fuzzy and medieval, it's not quite as bad as it sounds but it is still pretty vague.

HOLDING A CANDLE TO THE STARS

The theory goes something like this. If I take a bright object, a candle, say, and hold it up on a dark night, the farther away it is, the dimmer it gets. So if I have a way to measure the brightness of a candle I see at a distance, knowing how bright a candle is close up, I can work out how far away a distant light source is. We have very accurate instruments for measuring brightness: some can detect individual photons of light. (This isn't as impressive as it sounds. It only takes about a dozen photons to trigger the optical nerve in the human eye. On a clear, dark, unpolluted night a candle flame is visible to the naked eye fourteen kilometers away. But the photon detectors don't just see the light as our eyes do, they measure how often those photons arrive to give a clearer picture of brightness.)

So, if we knew how bright a particular star actually was, and could compare it with how bright it appeared to be, we could easily work out how far away it was. The catch is, we don't know how bright any particular star is. What makes it possible to use the standard candle theory is by assuming that there are some clearly identifiable types of star that are particularly consistent in their brightness. If we assume that all the stars of this class are the same brightness, then by identifying this particular type of star, and finding how bright it looks, we can work out its distance. This is one step better than Herschel's approach, which assumed all stars were equally bright, but it's still a big assumption. We don't know for certain that these types of star are all the same brightness. We just have to hope.

It might seem that finding out the type of star at a great distance is equally difficult, but there's plenty we can discover about the nature of stars despite their remoteness. We know what elements are in them, for example. The spectrum of light colors given off by stars has black gaps in it, corresponding to the energies of photons that are absorbed by the different elements that make up the star. Using spectroscopy, the technique of analyzing the colors emitted, and hence energies (see page 86 for more detail), it's possible to work out how the different stars are made up. This can be used to put stars into families with similar characteristics. Even better, there are some stars that also have very peculiar and particular habits that are easy to identify from a distance. One of the most commonly used standard candles is such a star, the Cepheid variable.

MISS LEAVITT'S VARIABLES

Variable stars, as the name suggests, fluctuate in brightness, pulsing in a regular fashion from darker to lighter in a period that can last anything from hours to a year or more. Cepheid variables are named after the constellation Cepheus. Anglo-Dutch amateur astronomer John Goodricke discovered the first true variable star, Delta Cephei (hence the name Cepheid) in 1784, two years before his death at age 21. He had already observed another star with varying intensity, Algol (Beta Persei), but this, as he suggested, varies because it is a pair of stars orbiting each other, one dark, not due to direct variation of the star itself. From observation of a good number of Cepheid variables that we can get a parallax distance on, it seems very likely that the speed of flashing of these variable stars is directly linked to their brightness.

The use of Cepheid variables as standard candles comes directly out of the work of one of the first prominent women in astronomy, Henrietta Swan Leavitt. Although Leavitt was never an observer, she was to spot the significance of this link between flashing speed and brightness. She was born in Lancaster, Massachusetts, in 1868, and as were a number of other women, she was first involved in science as a computer. This was not a device back then, as we now take "computer" to mean, but a person who made measurements and calculations from photographic plates. Despite ill health, she rose to become the head of stellar photometry at the prestigious Harvard College Observatory and was directly responsible for the theory linking speed of variation with brightness that would make Cepheid variables so valuable, thought up shortly before her death at the age of fifty-three in 1921.

Cepheid variables pulse over a period of days to months and it is believed that they are shrinking and growing to make those changes in brightness. It is thought that they don't have a proper balance between the two great forces on a star: gravity pulling inward and the pressure of the nuclear reaction that fuels the star pushing outward. In a Cepheid variable, gravity succeeds in pulling the star's mass inward; this increases the pressure, blowing it back out again. Then gravity takes charge once more, and so on. When it's contracting, it gives off less light, producing the trademark pulsation in its brightness. So finding a Cepheid in a distant location makes it fairly likely that we know how far away it is.

Although standard candles remain the workhorse of cosmological range finding, there is a more recent technique that should be able to give a more accurate measurement. When Einstein suggested that light would be bent by the distortion of space caused by a massive body, he also realized that such a heavy body can do exactly the same thing as a lens: bend the different light rays from a small source back together, so we get a clear picture of it, farther away than we could otherwise see.

Sometimes this gravitational lensing effect can result in our seeing more than one copy of the same astronomical body. Imagine a very bright, very distant object such as the incredibly bright proto-galaxies called quasars. As the light from them heads toward us across billions of light-years, some of it will pass by dense regions that bend the space through which the light is traveling. We can end up seeing more than one copy of the same body out in space. As quasars don't keep the same brightness, we can track that changing intensity for the pair through time.

What is found is that although they dim and brighten with the same rhythm, they are out of step because of the different route the light has taken. Although it is not trivial to account for the change in route, this means in principle that these data make it possible to work out how far away a quasar is without playing around with standard candles, and can even be used to help check the calibration of those candles.

But we've got ahead of ourselves. By the early nineteenth century, with Bessel's measurement of stellar parallax, we had some idea of the distance to the nearer stars, but no real concept of the size of the universe. Although the ancient Greek universe that was really more like the solar system had long been ditched, in the early twentieth century it was still assumed by most that the Milky Way—our galaxy, spanning around 100,000 light-years—was the whole universe. So we had gone from universe as solar system to universe as star cluster. But not everyone believed this.

ISLANDS IN SPACE

Remarkably it was William Herschel who was again ahead of the pack in suggesting that the Milky Way was just one of many galaxies. It has sometimes been said that Herschel believed that everything was inside the Milky Way, but this is totally incorrect, mixing together his observations on different types of nebulae, fuzzy patches of light in space. In the second of two papers called "On the Construction of the Heavens," Herschel asserted from his observations that the Milky Way was itself a nebula, like many others he had cataloged through his telescopes. In his growing number of catalogs, with his unparalleled telescopes, he was able to show that nebula after nebula was in fact a collection of stars, like our own Milky Way, but separated from us like islands in space.

Herschel had already decided there were a number of different types of galaxy (he still called them nebulae), some spherical, and some like the Milky Way (which he decided was what he called a Form III nebula) more like a flat pancake. Nebulae such as that in Orion he referred to as "telescopic Milky Ways," consisting of multitudes of stars. However, and this is where the confusion has arisen, he later pointed out that not every nebula was a vast cluster of stars. Some appeared to be a cloud of glowing dust around a star, and this type of nebula he decided (correctly) may be the birthplace of a star.

As well as missing Herschel's suggestion that the Milky Way was just one of many nebulae, many have also missed the significance of the approach Herschel took. Many years later, the great New Zealand–born British physicist Ernest Rutherford would comment, "All science is either physics or stamp collecting." Although this seems to have been partially said as a way of irritating other scientists (and is still very effective at upsetting biologists, who are more than a little touchy about the origins of their science), there was a point behind his comment.

Until relatively recently, most of science outside of physics—certainly most biology—was about cataloging and classification. It made no attempt to explain why or how things happened, it just detailed what was. Such information collecting is essential if you are to make the more sophisticated science of explanations work, yet it was probably fair that Rutherford labeled it "stamp collecting."

In Herschel's day, and for a good while afterward, practically all astronomy was science of the stamp collecting variety. Even Herschel himself did a vast amount of cataloging, and is most famous for doing just that, particularly when he added Uranus to the small list of planets known since ancient times. However, when he was writing about nebulae, Herschel did much more. He tried to explain how nebulae had come into being, to give a description of the processes behind what was now observed, rather than simply reporting what he saw, and this was quite revolutionary. He had moved astronomy into the more advanced phase of science.

Herschel's theories caused a degree of upset, as his model of nebulae formation required gradual gravitational attraction, implying that the universe had taken a long, long time to come into the form it now had, whereas at the time the assumption was that it was created in a handful of days as it said in the Bible. It was thought that this was something best left to theologians, rather than allowing scientists to trample into the arena. This may provide a reason for why Herschel's ideas are not now widely known: many in the astronomical hierarchy of the time were clergymen who held a dim view of his theories and preferred to ignore them.

Herschel's ideas were all but forgotten. We can see this from a popular book on the universe written in 1905. Called *The System of the Stars*, it was written by Agnes Clerke, a distinguished Anglo-Irish astronomer and writer. In it, she commented, "The question whether nebulae are external galaxies hardly any longer needs discussion. . . . No competent thinker, with the whole of the available evidence before him, can now, it is safe to say, maintain any single nebula to be a star system of co-ordinate rank with the Milky Way." Dangerous words.

IMPROVING TECHNOLOGY

Over the years from Herschel's first attempts at relative measurements, telescopes got better and better. Although Herschel's own giant telescope with a 1.2-meter (49-inch) mirror had been something of a failure, difficult to steer and harder still to observe from, it was by no means the last word in expansion in the optical world. Over in Ireland, the eccentric third Earl of Ross built the Victorian observing masterpiece nicknamed the Leviathan of Parsonstown (now Birr) where Ross' estate lay. This instrument had a 1.8-meter (72-inch) mirror and was more successful than Herschel's great telescope, in part because it was more limited.

To enable it to point anywhere in the night sky, Herschel's instrument had been mounted on an enormous rotating wooden structure that took a huge amount of effort to drag into place, and lacked the absolute stability required for clear images. Ross' Leviathan, by contrast, was set between two stone walls. This meant it had a very restricted field of vision, but made it much more solid and easy to swing into position. The Leviathan would be trained on the skies from 1845 right through into the early twentieth century, although its use was always dogged by the wet and misty Irish weather, leaving relatively few clear observing nights in any given year. Unlike Herschel's telescope, which was dismantled, the Leviathan has been restored and can still be seen at Birr in all its glory.

Other telescopes would be built in Europe, but by the time the Leviathan was pensioned off, the baton of astronomical greatness had shifted to the United States, where better weather conditions and the kind of money required for increasingly expensive astronomical projects were more readily available. In 1908, the Carnegie Foundation offered funds for a large telescope to be built for the California Institute of Technology under the guidance of astronomy professor George Ellery Hale.

Hale, born in Chicago in 1868, attended MIT and went on to work at observatories from Harvard to Chicago before becoming America's preeminent founder of new observatories. Two mountaintop sites were considered by Hale for the new Carnegie-funded giant: Mount Wilson near Pasadena and Mount Palomar in San Diego County, both in California.

Each offered locations high enough to reduce the distortions caused by the atmosphere. Palomar was better in terms of light pollution (it was in the middle of nowhere, whereas Wilson was near Los Angeles) but Wilson had better access and so it was chosen for Hale's 1.5-meter (60-inch) device. This was smaller than the Irish telescope, but had a much better mirror (silvered glass instead of copper/tin alloy), worked under much clearer skies, and had the ability to steer in any direction, giving far better results than the Leviathan ever had. It was augmented in 1917 by the 2.5-meter (100-inch) telescope, which was the largest in the world for thirty years, and would play a crucial part in the next step of understanding the scale of the universe.

Even with Lord Ross' Leviathan, and more so as the American telescopes came on line, it had been obvious that some nebulae had complex structures and appeared to incorporate stars as well as gas and dust. Herschel had seen this clearly enough. This didn't rule out their being within the Milky Way, but gave more credence to the island galaxies theory.

THE GREAT DEBATE

By the 1920s, the issue of what the Milky Way and the nebulae were had still not been settled. One particular debate on the subject in April 1920 would inspire the quirky American astronomer Edwin Hubble to transform our view of the universe.

Such debates themselves had been going on for a long time, certainly ever since Herschel's day. To the naked eye, the few nebulae that are visible look like slightly fuzzy stars, but once telescopes of any power were available, it become obvious that they had a more complicated structure, some near-spherical, others with a tail like a comet but extending in two directions, and the most dramatic spiraling around the center, forming a whirlpool-like structure.

The prevailing idea was still that nebulae were young stars, in the process of forming. Stars form by the gradual accumulation of matter. Gravity pulls together bits and pieces, mostly hydrogen, floating in space. When some has accumulated, it has a bigger gravitational pull, so it attracts more matter. Eventually this ball of matter is so dense that nuclear fusion (see page 102) begins. It seemed reasonable that before a star was fully established there should be clouds of material that are on the way to being a star, and this is what it was thought nebulae were. (This wasn't so stupid: some nebulae are indeed stars forming in our galaxy.)

However, Herschel's alternative viewpoint, that nebulae were "island universes," other collections of stars like the Milky Way, but immensely distant, had also lasted through to the 1920s, boosted by support from his older contemporary, the philosopher Immanuel Kant, who seems to have independently come up with the idea. This would mean that the Milky Way was not the whole universe, but just a single one of the nebulae (what we would now call galaxies), nebulae which were spread across a vast expanse of space in the same way that the stars are spread through the Milky Way. If Herschel and Kant were right, the tails that appeared to come out of some nebulae were the effect of seeing a disk sideways on.

The formal debate took place in April 1920, organized by the National Academy of Sciences in Washington, between the Mount Wilson astronomers, who stuck with the more popular "Milky Way is everything" theory, and competitors from the University of California's Lick Observatory on Mount Hamilton who believed our galaxy was just a small portion of a vastly bigger universe. In the Wilson corner was Harlow Shapley, and for Lick, Heber Curtis, both well-established astronomers (Shapley was in his mid-thirties and Curtis in his forties), presenting strongly held beliefs in a debate that was not made easier when the two men had to spend many hours beforehand on the same train from California. Each put forward his case, and neither really succeeded in changing many minds.

Although there was interesting evidence of complexity in the nebulae coming from the new telescopes, there were also arguments that seemed to carry weight against nebulae being anything more than embryonic stars. They always seemed to appear where you would expect them to be if they were local gas clouds, and one had recently become an extra one-tenth as bright as before when a nova flared up, something that seemed out of proportion if a nebula really was a vast distant collection of stars. By coincidence, this was the Andromeda nebula, which later would have a significant impact on the revelation from Mount Wilson that established the scale of the universe.

Although most of those at the Mount Wilson Observatory still believed that the Milky Way was the entire universe, there was a significant dissenting voice—like Herschel, a great anglophile—Edwin Powell Hubble. At the time of the great debate he was thirty-one, solidly established in astronomical circles, but yet to make his mark in a big way.

THE MAKING OF HUBBLE

Born in Marshfield, Missouri, in 1889, Young Edwin had been introduced to astronomy when his grandfather made him a telescope as an eighth-birthday present. His immediate delight in what he could see through the crude instrument blossomed rapidly into an out-and-out passion. Edwin Hubble, even at this early age, was someone who fell in love with a concept and stuck with it through thick and thin.

It was only Hubble's father who couldn't see that his son was destined for a sparkling career in astronomy. Rather like the parent of a would-be actor, Hubble's dad was determined that his son have a "real" career as a fallback in case this study came to nothing and insisted that, when he went to the University of Chicago, Hubble specialize in law. Hubble senior was really only reflecting the realities with which he was familiar. Until the twentieth century, most astronomers were amateurs, wealthy individuals who studied it for fun, rather than treating the subject as a serious career. Even today, astronomy is one of the few fields of science where amateurs can still make a significant discovery.

With a combination of political savvy and sheer determination, Hubble managed to undertake his law degree while also attending classes that would contribute to the future in astronomy that he desired. It was ironic that although Hubble had no problems with his law work, and would have made a passable lawyer, it was his sideline in science that got him put forward for a select academic prize to study law further, a Rhodes scholarship.

Set up in 1903 in the will of the phenomenally powerful Cecil Rhodes, the founder of both the de Beer diamond mines and the country of Rhodesia which later split to become Zimbabwe and Zambia, the scholarships provide places at the University of Oxford for thirty-two Americans each year. (There are also scholarships for a range of other ex–British colonies and now, although not in Hubble's day, Germany.) This opportunity to round out the education is usually chosen for individuals who are expected to go far, potential movers and shakers; other Rhodes scholars have included leading congressmen, Supreme Court justices, senators, and state governors, including former president Bill Clinton.

Thanks to the recommendations of Nobel Prize winner Robert Millikan, Hubble got a scholarship and traveled to Oxford in the fall of 1910, still technically tied to the law, although he also touched on astronomy when he could. It was at Oxford that Hubble underwent an extraordinary transformation, consciously taking on the mantle of the kind of British gentleman that would later be portrayed in Hollywood when talking pictures came along, adopting an artificially clipped English accent and making interjections of the "What ho!" variety worthy of P. G. Wodehouse. He indulged in a heady taste for tweed clothes and took up a briar pipe with great enthusiasm. Just as he had with astronomy, Hubble fixed on something that delighted him and hung on to it with the tenacity of a terrier.

The chances are that Hubble would have stayed in England had it not been for personal circumstances. In January of 1913, soon after his Rhodes scholarship had expired, he had to hurry back home to try to salvage the wreckage of his family's finances when his father died. He took on what work he could, mostly in teaching, until the family fortunes were reestablished, then finally out from under his father's yoke, he was able to abandon any thought of the law and plunge into full-time astronomy.

A REAL ASTRONOMER

His passion for the subject first took Hubble to the Yerkes Observatory at the University of Chicago, where he developed an interest in nebulae. These fuzzy patches in the sky required as much magnification as you could get to see them clearly. The Yerkes had the world's largest functioning refractor, a traditional, lens-based telescope, rather than one like Mount Wilson or Mount Palomar, which use a mirror to collect and focus the light rays. At 1 meter (40 inches) across this was big indeed, but couldn't compare with the best new telescopes. In 1919, after a time in the armed forces toward the end of the First World War (followed by a tour of England) he took up a position at the Mount Wilson Observatory where the 1.5- and 2.5-meter giants offered him the sheer power that he needed.

The life of the astronomers based at Mount Wilson in the 1920s was a strange one. Not only was the site isolated, high up a narrow track littered with hairpin bends, but working there involved a rigorous schedule where nights on the telescope were interlaced with days interpreting the photographic plates used to record images of the stars. In the basic accommodation up on the mountaintop that they called the Monastery, there were only limited comforts, and the observatories had to be kept at freezing night-time temperatures to avoid distortion from the haze any heating would cause. Hubble took to striding about the place in his army greatcoat, cementing his eccentric image.

The breakthrough in our understanding of the scale of the universe came in October 1923 when Hubble discovered a Cepheid variable in what was then known as the spiral nebula in Andromeda, or M31 from its Messier catalog number. (French astronomer Charles Messier cataloged over a hundred nebulae and star clusters in the 1770s because he was frustrated by frequent reports that mistook nebulae for new comets. Many of the most visible nebulae are still often known by their number in Messier's catalog.)

We now know M31 is the nearest spiral galaxy to our own, but at the time it was argued that it might be just a luminous cloud in the Milky Way. Using the 100-inch telescope, the only one in the world capable of detecting Cepheids this far away, Hubble found that the timing of the Cepheid in Andromeda carried a shocking message. He calculated that its period of just over a month made its distance 900,000 light-years away, far beyond the boundaries of the Milky Way. Although a member of the Mount Wilson group that had championed the "Milky Way as universe" theory, he had proved his colleagues wrong.

A TRANSFORMATION OF SCALE

Hubble's measurements suggested that the traditional Mount Wilson view vastly underestimated the scale of the universe. Ours was but one of many galaxies in a huge, and possibly infinite, expanse of space. Some scientists, particularly in the media-frenzied twenty-first century, might be inclined, having made such a discovery, to rush into a press conference to announce this remarkable finding to the world. Let's be clear what Hubble had done. Before his work, the consensus was that the universe was the Milky Way, a disk around 100,000 light-years across. Hubble had discovered that one of the nebulae was nearly ten times this distance away, and this was the brightest one in the sky, visible to the naked eye. It was quite possible that others were much more distant. The universe had suddenly vastly transformed in scale.

However, as a good scientist should, rather than blurt out his results straightaway, Hubble went on to take more photographs of the Andromeda nebula, looking for further confirmation of his findings, and discovering a second Cepheid variable that backed up his results. Three months after first spotting the landmark star, he went public. It might seem this reluctance to blow his own trumpet was a reflection of the stiff upper lip that went with his adopted

British character, but Hubble was no shrinking violet; he enjoyed showmanship and being center stage. It's more likely that it was because, bearing in mind Mount Wilson's reputation as the spiritual home of the Milky Way as universe theory, he knew he had to be 100 percent certain before going against the Mount Wilson "family values."

Even more remarkably, Hubble found evidence that supported a theory that had been proposed by his contemporary, the Dutch astronomer Willem de Sitter. This theory suggested that the universe was not fixed in size, but was constantly expanding. This revelation from Hubble we come back to in Chapter 5 when we explore the gradual birth of the Big Bang concept. For the moment, Hubble had given our first real insights into the size of the universe.

By the mid-1930s, Hubble was prepared to make significantly more detailed pronouncements on the natural history of the universe itself. In an address to the National Academy of Sciences in Washington in April 1934, Hubble made some impressively sweeping statements. He said that the universe was "a finite sphere" 6 billion light-years across, and that it was made up of 500 trillion nebulae, "each unit being 80 million times as bright as the Sun and 800 million times as massive."

Inevitably some of Hubble's statements were little more than guesswork. The number of galaxies, for example, was simply plucked out of the air given how frequently they seem to appear locally and the assumed size of the universe. But there was a much more essential component of Hubble's work that relied on something equally tenuous. He had managed to measure the distance to the galaxy in Andromeda using Cepheid variables, but this technique was useless for more distant galaxies because there weren't bright enough Cepheids to detect from that range. Instead, he had to come up with a new standard candle that by comparison made the use of Cepheids seem rock solid; he reverted to a more sophisticated version of Herschel's siriometer.

STAR LIGHT, STAR BRIGHT

Inevitably, the stars you could see in a distant galaxy were the brightest ones. That was indubitably true. So Hubble made a big leap of faith. What if the brightest star in any particular galaxy was of a similar brightness to the brightest star in any other galaxy?

Such a technique, which assumes that the extremes of a population are similar, is a dangerous one in some circumstances. It doesn't work if the structure of the items you are comparing is very different. So, for example, there is little resemblance between the heights of the tallest mountain in the United Kingdom, Ben Nevis at 1,344 meters (4,400 feet) and the tallest mountain in the United States, Mount McKinley at 6,194 meters (20,320 feet).

Nor does drawing such parallels work with small populations, where you might see a big difference between two samples within the population. If I pick one hundred people at random from the sidewalk of New York, the chances are that their ethnicity won't be representative of the population of the United States as a whole. But where there's a big population of similar structures, then the comparison becomes more realistic. If we look at the ages of the oldest person in Europe and the oldest person in the United States, then they are likely to be quite similar; they certainly won't vary the way the mountain heights do.

Galaxies have huge populations of stars, so assuming that we're dealing with similar kinds of galaxies, then it isn't entirely unreasonable to assume that the brightest stars would be similar in intensity. It was relatively easy to spot physically similar galax-ies, although it wasn't always so easy to deal with age (which could also influence how bright the brightest star is). Even so, Hubble had found a yardstick that would enable him to make best guesses of measurements well beyond the detection distance of Cepheid variables.

AS FAR AS THE EYE CAN SEE

All the attempts we've seen to measure the size of the universe only refer to the visible size. If it has a finite age (of which more in a moment), then we can only ever see as far as light can travel in the time available, a timescale that is now generally thought to be around 13.7 billion years. So the biggest extent of the universe we can see at the moment is around 27 billion light-years across. (In a billion years' time, that expanse will be 29 billion light-years, and so on.) But that doesn't mean the universe stops at the limits of what it's possible to see.

Although this observed universe is inevitably finite, the true universe could still be infinite. In fact having an infinite universe is in some ways easier to handle. If the universe has a boundary, you always have the problem of what's beyond it. If, as we believe, our universe is expanding, into what is it expanding? That's rela-tively easy: it's quite possible that it is nothing. The theories of an expanding universe say that space itself is expanding, so there doesn't need to be any space into which it expands.

A more tricky question, however, is if the universe is finite, what happens at the edge. If you could fly a spaceship into the edge of the universe, what would happen? If space literally stopped, then it is possible there could be a barrier. If that were the case it would be as if you hit an invisible wall. There just would be no space to go into, so you couldn't go any farther. (In practice, such an experiment may well not be practical to try, as the universe's expansion probably means that the universe is getting bigger at a faster rate than we could ever travel through it.)

An alternative view that has frequently resurfaced over the years is that the universe folds in on itself in a fourth spatial dimension. This theory suggests that if you flew off, apparently through the edge of the universe, then you would fly back in the opposite side. From your viewpoint, there would be no edge, the universe would go on forever, but you would eventually pass the same point over and over again.

This isn't as bizarre a picture as it seems. We can imagine equivalent situations for lesser-dimensional beings without any problem. It's the sort of thing a one-dimensional being would experience following a circular line, or a flat being would experience going around the surface of a sphere. You are moving in what appears to you to be a straight line, your journey never ends, but you keep coming back to the same spot. Your world is finite, you can travel all the way through it, yet it is unbounded. You never come to an end. The ancient Greeks were wrong when they thought you couldn't have something finite that didn't have boundaries.

AN ORIGAMI UNIVERSE

Cosmologists now have a map of the early universe to play with, which shows the distribution of something called the cosmic microwave background radiation. This is a map of the pattern of light emerging from what is believed to be the early universe. If this idea of a finite unbounded universe is true, you might expect the light that was heading out of the universe in one direction to be visible entering it from another. Inevitably the pictures we have from the WMAP (Wilkinson Microwave Anisotropy Probe) satellite that show this background radiation are something of a fuzzy mess of detail, but it was hoped if scientists only took an overview, or concentrated on the higher-intensity sources, they might be able to see some visual reflections of one part in another, caused by the image passing out of the universe on one side and back in the other.

It didn't work. They couldn't see an appropriate repetition of images. However, in 2007 Boudewijn Roukema and colleagues at Nicolaus Copernicus University in Poland came up with a subtle twist on this idea that does seem to be backed up by the evidence. We tend to assume that the universe is spherical. This goes back to the sort of reasoning Roger Bacon described, that anything other than a sphere would have problems if it rotated because parts of it would be moving in and out of nothingness.

However, if we assume the universe is a dodecahedron (a twelve-sided solid where each side is a pentagon) then an image coming out one side and back in another would be rotated through thirty-six degrees, reflecting the difference in orientation of the two sides. Roukema found that if you cut rings out of the WMAP picture of the early universe and rotated them through thirty-six degrees, then it was possible to match up these rings with equivalent unrotated rings in the "opposite" side of the universe.

As yet we don't have good enough data to show that this effect is definitely happening. Until there is a more detailed map of the cosmic microwave background, which it is hoped will be obtained from the European Planck satellite (launched in May 2009) over the next few years, there remains a roughly 1 in 10 chance that this apparent match is just random coincidence. But there is still some evidence here of a finite unbounded structure, providing a fascinating insight into the possible form of the entire universe.

We are increasingly sure we know at least part of the story on the size and large-scale components of the cosmos. Yet this alone doesn't give a good enough picture of what's out there. We also need to know more about the universe's age.

4.

HOW OLD?

> I came into the room, which was half dark, and presently spotted
> Lord Kelvin in the audience and realized I was in for trouble at the
> last part of my speech dealing with the age of the earth, where my
> views conflicted with his. . . . Then a sudden inspiration came, and
> I said Lord Kelvin had limited the age of the earth, *provided no new*
> *source was discovered.* . . . The old boy beamed on me.
>
> —ERNEST RUTHERFORD (1871–1937),
> quoted in *Rutherford, Being the Life and Letters*
> *of the Rt. Hon. Lord Rutherford* (A. S. Eve)

In ancient times the age of the universe was set by decree. Many of the ancient Greek philosophers believed the universe had gone on forever; it had no beginning and would have no end. Aristotle, in talking about infinity, considered that time had always been ticking away, so the universe had to be without a start point. After all, if it did start, what came before it?

However, although the philosophy of the Greeks was much revered by medieval scholars, this aspect of Greek thinking was considered flawed, because for those adhering to many religions, including those that arose from the Middle East that would come to dominate the Western world, there was a clear starting point for the universe, described in their sacred texts.

USSHER'S CALCULATION

Famously, biblical scholars undertook complex calculations based on the genealogies in the Bible, working back from historical times to creation, but these figures require a degree of interpretation, so figures were produced for the age of the universe that ranged from 6,000 to 9,000 years. The best-known date of this kind, that would come to dominate the field, was that produced by the Irish-Anglican bishop James Ussher.

Ussher pronounced that the world was created in 4004 BC. Not to be accused of inaccuracy, he pinned down creation to nightfall on the evening before Sunday, October 23, 4004 BC (that's September 21 in the modern calendar, as they were still using the Roman calendar in Ussher's time, which didn't properly handle leap years). This date was accepted as fact by the Anglican Church and others in the eighteenth century. So well established was the date back then that it was printed in the margin in Genesis in the King James Bible, and would remain there right up to the early 1900s.

However, the existence of this date didn't stop scientists thinking about a wider range of possibilities. It is often said that the 4004 BC date was first questioned when Darwin's theory of evolution began to cast doubts on the literal nature of the Bible's genealogy, or when geology first began to provide indications of the great age required to lay down the geological strata, but in fact there was scientific evidence significantly predating these developments.

HERSCHEL'S LAST SURPRISE

As we have already seen, the great German/British astronomer William Herschel had made an estimate of the size of the Milky Way that made it clear that the universe had to be quite old. He was aware that the time light took to arrive from the stars made it obvious that he was peering back much farther than 6,000 years. Herschel went even further, suggesting that the universe was millions of years old, but frustratingly we don't know on what evidence this statement was based. This was as early as 1813, long before anyone else seems to have dared to suggest it.

On September 15, 1813, the Scottish poet Thomas Campbell wrote to "a friend" about his recent visit to that "great, simple, good old man" William Herschel (Herschel was seventy-five at the time). According to Campbell, Herschel said this (the italics are from the original source):

> I have *looked further into space than ever human being did before me.* I have observed stars of which the light, it can be proved, must take two million years to reach the earth.

This was a remarkable observation as it smashed to pieces any possibility that the universe had only been created 6,000 years earlier, yet Herschel's insight seems to have largely been ignored or forgotten, even by those chronicling the scientific understanding of the age of the universe.

CATASTROPHIES AND GRADUAL CHANGE

Herschel may have been ahead of his times, but over the next hundred years, the most important change for forming a sensible view of the age of the universe was not so much a specific theory, but a different outlook on the kind of actions that could have produced the present form of the universe and the Earth. The Bible describes very sudden changes. According to Genesis we went from there being nothing to the Earth being present, roughly as it is now, in a day. The surface of the Earth would be drastically changed again in the great flood. These were catastrophic changes, not gradual evolutionary change.

Because of this, the assumption had been that what we see around us tended to have arisen in huge, sudden, and dramatic ways. But science began to pick up suggestions that gradual change was more common in natural processes. When scientists first began to develop an explanation of how sedimentary rock was laid down, they discovered not a sudden dramatic change but a very gradual one as layer after layer was added. Charles Darwin, as much a geologist as a biologist in his early work, had this same gradual approach in mind when he described his theory of evolution through natural selection. This was not the sudden dramatic introduction of species but a very slow, piece-by-piece change over thousands or millions of years.

This change in attitude to the way the Earth and the universe came to their present form would also have a strong influence on cosmological ideas. Where once beginnings were seen as catastrophic events, they were now considered much more uniform, slow and steady. Oddly enough, although this "uniformitarian" approach has dominated science ever since, recent understanding of the way complex systems change, typified by Malcolm Gladwell as "the tipping point" and also seen in the butterfly effect from chaos theory, has led us to realize that whereas much change is gradual and uniform, it is very easy to get a complex system into a state where a very small initial modification can result in a very large and rapidly changing outcome.

WHICH CAME FIRST, THE EARTH OR THE UNIVERSE?

Because of the difficulty of pinning down cosmological facts, it has frequently been the case that the dating of the age of the universe has been out of step with the dating of the age of the Earth. The task is much easier for geologists, who can get their hands on the materials around us to date them, than it is for cosmologists. This led to several embarrassing periods when the best estimate of the age of the universe was lower than the best estimate of the age of the Earth.

Initially this wasn't a problem, because with the exception of Herschel's prescient observation, there were very little data from astronomical sources that could be used to apply an age to the universe, and there was more and more evidence coming along that gave a minimum age for the Earth. Although this was the case, it was easy enough to say that, however old the Earth was, the universe had to be at least as old. When it was decided from the study of geology that the Earth must have been around at least a billion years, then it was easy to say the universe went back that far too.

Even working on something less grand than the age of the entire universe, there have been dating inconsistencies. Around the end of the nineteenth century, when it was becoming obvious that the Earth was a lot older than first thought, there was no sensible explanation for how the Sun shone, because nuclear fusion simply didn't exist as a concept. Assuming that it worked by burning material, like an immense coal fire, the Sun was given a maximum lifespan of only a few million years, far too young to be the home star of a much older Earth.

A good example of a point in time when the age of the whole universe appeared to be younger than that of the Earth was when astronomer Sir Arthur Eddington gave a lecture on the expanding universe to the British Association for the Advancement of Science in September 1933. Eddington, whom we will meet a number of times, was a vastly influential astronomer and supporter of Einstein on relativity. He was among the founding fathers of the discipline combining astronomy and physics, astrophysics. Born in Kendal in England's Lake District in 1882, Eddington was educated at Manchester and Cambridge, where he would remain for his working life except for a brief period based at the Greenwich Observatory. He was also one of the first great popularizers of science, writing for and lecturing to the general public.

According to Eddington in 1933, when the age of the universe was calculated by extrapolating the current expansion of the galaxies away from each other back over time, the result was a figure of "not more than 2,000,000,000 years ago," less than two billion years at a time when the Earth was known to be over three billion years old. But, as Eddington pointed out, "This assumed that the velocities of the systems of nebulae had always remained of the same size, but if one allowed for the gravitational attraction, for the going apart against the attraction with velocities greater in the past than now, that limited the time still more to, say, [one billion] years."

In other words, if we assume that the expansion of the universe is slowing down, which it seemed likely to do because the galaxies exert a gravitational pull on each other, then it must have flown apart from the beginning even faster than it is now, so that means that it's even younger than we think. The crisis in aging when compared with the Earth would be even greater.

By 1933, an early version of what would eventually be called "dark energy" (see page 166) was being put about. This "cosmic repulsion" would cause the galaxies to accelerate away from each other, meaning they started off slower, and so gave the universe longer to come into being. Eddington reckoned this would make it possible for the universe to be up to ten billion years old, which gave time for the Earth to evolve, but some astrophysicists doubted it was enough time for all the stars to come into being, and at the time this cosmic repulsion was considered something of a crank concept.

RR LYRAE TO THE RESCUE

By the mid-1940s there was still a widely held view of the age of the universe that made it around half as old as the Earth, not a sustainable position. Strangely, the first sign of at least part of what was wrong with the accepted age of the cosmos emerged from a study that was setting out to confirm the size of the universe. Remember those Cepheid variable stars, used as standard candles to measure distance. Because there was a big assumption in using standard candles, the ideal would be to find another measure against which to check the Cepheid variables. There were other types of variable star, and it was hoped that by finding them in the same galaxy as Cepheids it would be possible to help confirm the distances involved.

Although having confirmation wouldn't overcome the central "cross your fingers and hope" assumption that variable stars of the same frequency would be equally bright, it should at least give extra weight to the figures. The type of stars selected was called RR Lyrae (again after the location in which they were first identified) variables. There is a whole range of variable stars, some of which vary for different reasons from the swelling and deflating Cepheids, and others with very different or irregular pulsations. The RR Lyrae variables are very similar to the Cepheids, but significantly weaker.

This strange name "RR Lyrae," sounding more like a steamship than a heavenly body, refers to a particular star. Traditionally stars in a constellation are named using their "Bayer designation," an approach named after the seventeenth-century German astronomer Johann Bayer. He started giving stars names that used a letter of the Greek alphabet followed (confusingly) by the genitive Latin version of the constellation.

So, for example, the constellation of the Centaur became Centaurus in Latin. The genitive (meaning "of the Centaur") is Centauri, and the first star is α Centauri (Alpha Centauri), which happens to be the closest star to the Earth apart from the Sun. (Just to complicate matters, Alpha Centauri is actually three stars that orbit each other, and it's the faintest of the three, Proxima Centauri, that is specifically the closest to us.) Exactly how "first" is determined within a constellation is rather messy. Although the order is supposed to go by brightness, this wasn't well established back in the 1600s and Bayer often combined approximate brightness with location.

The names went on through the Greek alphabet, then through the lowercase modern alphabet. On the odd occasion he needed more, Bayer included the uppercase modern alphabet, but never got past Q. When it came to naming variable stars (assuming they hadn't already got a name) they started from R through to Z, then RR to RZ, SS to SZ, and so on. After then going through AA to QZ (omitting the Js which are too like I) they then switched to a more modern convention using V for variable, followed by a number that started at 335 as there had already been 334 letter variants. No one said astronomers like to make things simple.

So RR Lyrae is the tenth variable star in the constellation area of Lyra, but it forms the definitive star for this particular type of variable. Although RR Lyrae variables are similar to Cepheids they can be distinguished as the stars are less massive with shorter pulsations. It's just as well there is this distinction, or it wouldn't be possible to tell whether a weaker star was a closer RR Lyrae or a more distant Cepheid. By the mid-1940s, no RR Lyrae variables had yet been seen in another galaxy, even one as relatively close as the galaxy in Andromeda.

THE CASE OF THE MISSING STAR

However, astronomy is always improving its tools, and in 1948, the U.S. triumph of Mount Wilson's 2.5-meter (100-inch) telescope was bettered by the new instrument at Mount Palomar also constructed for the California Institute of Technology, and once more under the aegis of the Carnegie Foundation and masterminded by George Hale, after whom the telescope was eventually named. This had a mighty 5-meter (200-inch) mirror. What's more, the new telescope was farther away from the limiting glow of cities, giving it a huge lead over its older relative.

This was a major achievement for America. When I was first interested in astronomy in the 1960s, Palomar's telescope was still the most powerful in the world, a beautiful monster, the greatest ever of the traditional giant instruments, that would only be beaten by switching to new technologies made possible as much by computerization as by optical developments. The sheer size of the mirror, made from sixty-five tons of glass, is phenomenal. It took around fifteen years to build, interrupted by the Second World War, and by the time it was inaugurated in 1948, with the mirror and telescope supported by a five-hundred-ton mount, Hale had already been dead for ten years.

Astronomer Walter Baade was already an expert in RR Lyrae variable stars when he became one of the first to make regular use of the mighty Hale telescope at Mount Palomar. Baade was born in Schröttinghausen, Germany, in 1893; he emigrated to the United States in 1931 and made it his home for the rest of his working life. He had the bright idea of checking out Hubble's distance data, on which the age of the universe hung, by finding RR Lyrae variables in the Andromeda galaxy and measuring their periods—the rate at which the star's brightness varies. So far, so good. But search as he might, Baade couldn't find any.

Outside of science, this kind of failure could well be regarded as a disaster—and even in the scientific community, the failure of an experiment can wreck a career—but sometimes failure to find something provides useful data in itself. For example, when the American physicists Michelson and Morley failed to find any evidence for the existence of the ether, they triggered the chain of ideas that would lead to Einstein's special relativity. When Baade didn't find RR Lyrae variables in the galaxy in Andromeda, he too came up with a conclusion that would change our scientific outlook.

It's possible, of course, that there was something special about the Andromeda galaxy that meant it didn't have any RR Lyrae variables. But it had Cepheids, as Hubble's work had shown, and there was nothing about it that suggested it was much different from our own galaxy. There was only one other obvious possibility. That the RR Lyrae variables were there, but that Baade couldn't see them.

Given the relative strengths of the Hale telescope and its predecessors, it should have been able to pick up those variable stars. So unless the whole of the Andromeda galaxy were shrouded by a dimming cloud of dust (something that didn't seem to be the case from other observations) the most likely possibility was that the Andromeda galaxy was farther away than Hubble had thought. Much farther.

THE WRONG KIND OF CEPHEID

Hubble's first suggestion for the distance to the Andromeda galaxy, using Cepheid standard candles, had been 900,000 light-years, which he later revised down a little. If Baade was right, that 900,000 light-years was significantly too small. He went back to Hubble's original assumptions and there he found a critical error that had gone unspotted, an error based on knowledge that wasn't available when Hubble made his measurements.

The whole idea of using Cepheids as standard candles rests on the fragile assumption that all Cepheid variables are pretty much the same, and that their brightness can be deduced by how frequently they pulse. However, since the 1920s, astronomers had got a better understanding of the natural history of stars. Stars form from the coalescence of gas and dust, pulled together by the increasing gravitational pull of the body as it gets larger and larger. They are made up of whatever debris surrounds the area where they form.

In the early days of the universe, there was very little else but the hydrogen and helium (and much smaller amounts of lithium and beryllium) that are thought to have come out of the Big Bang and the post–Big Bang fireball. So the very oldest stars haven't got much else in them. Younger stars, such as the Sun, however, were formed after the heavier elements had been forged in the first generation of stars and spread throughout the universe after the explosions of supernovae. So these younger stars like the Sun have significantly more of the heavier elements in them. Such stars burn brighter and are known as population I, whereas the older stars with much less of the heavy elements are called population II.

Although there are degrees of variation, there is a clear difference in brightness between similar stars from population I and population II, and this applies to Cepheid variables just as much as it applies to any other star. The mistake Hubble made, as Baade discovered, was that he assumed that the Cepheid variables he had seen in the Andromeda galaxy were the same as the dimmer, older population II variable stars that had first been used to calibrate the standard candles that were first found in the Milky Way. But in fact they were population I stars, significantly brighter than he had assumed. This brightness made them easier for Hubble to pick them out, but misled him over the distances involved.

Once astronomers were aware of the differences between population I and population II stars, it was easy enough to tell the difference from spectroscopic analysis. So now Baade began checking Cepheid variables, and sure enough they fell into two clear classes, one around four times as bright as the other.

The brightness of light is a so-called inverse square law, which drops off as the square of the distance you are away from the source. It's easy to see why. From high school geometry, you might remember that the surface area of a sphere is given by $4\pi r^2$; it depends on the square of the sphere's radius. So the area around a star, which is how much the light is spread out, goes up as the square of the radius, the distance away from the source. If a star is four times brighter than you first thought it was, it is twice as far away as it first appeared.

With this new realization, it seemed that the galaxy in Andromeda was twice as far away, at 1.8 million light-years distant. And this was the yardstick by which other galactic measures had been established. So the whole universe suddenly grew by a factor of two. Ignoring Eddington's cosmic repulsion, as most astronomers still were (it was, after all, a truly bizarre concept), this meant the universe could be as old as 3.6 billion years. This still made the Earth surprisingly ancient in universal terms (it was thought at the time to be around 3 billion years old) but at least it had removed the huge embarrassment of the child being older than the parent.

FINDING THE EARTH'S BIRTHDAY

As it happens the age of the Earth was also inaccurately measured at the time. Geologists might have it easier than astronomers as they can touch and directly study the Earth, but they still had a nontrivial task in putting an accurate age even on specific rocks, let alone the whole planet. It required a breakthrough from physics to be able to provide accurate measurement of age.

The man behind it was one of the great names of experimental physics, Ernest Rutherford. Rutherford, who we've already met through his "stamp collecting" jibe, was born in Nelson, New Zealand, in 1871 of Scottish and English parents, and spent most of his working life at the universities of Manchester and Cambridge in the United Kingdom. But at the turn of the century he was working at McGill University in Montreal, and in 1902, Rutherford and a colleague Frederick Soddy wrote a paper suggesting that the newly discovered idea of radioactive decay could be used to tell how old rocks were.

Marie Curie in France and others had discovered that different materials changed their structure over time. Uranium, for example, gradually underwent a radioactive decay that turned it into lead. The rate of transformation was steady, so it could act as a historical clock. If you had a rock that had some uranium in it when it was formed, you could estimate how old it was by seeing how much of the uranium had been transformed into lead.

There were a number of problems along the way to accurate dating. Looking into the past required assumptions about how radioactive materials had been present in the past. Some of the short-term dating using radioactive carbon, for instance, was only made accurate once it was pulled into line with tree rings in ancient trees. But now we can say with some certainty that the Earth is 4.53 billion years old. This also seems to apply to the rest of the solar system. Nothing older has yet been found.

CHECKS AND BALANCES

By modern dating of the Earth, Baade's age for the universe was still impossibly young. But it wasn't to remain so, thanks to another important tenet of science. One-off observations or measurements are valuable for flagging an issue, but they can never be considered definitive. For an experiment to be accepted it needs to be reproducible. Someone else should be able to get the same results in a different laboratory. Similarly, an astronomical observation has to be repeated, ideally by someone else with a different viewpoint, to make sure it is up to scratch.

With this in mind, Baade set his assistant, Allan Sandage, to check his measurements using new observations, and this was to upset the age of the universe once more. Sandage didn't find anything wrong with Baade's measurements on the Andromeda galaxy, but rather in the way those measurements were then spread to more distant galaxies to work out the scale of the universe and hence to be able to track it back in time to its beginning.

Remember that Hubble had used the only available way to put a range on distant galaxies. Cepheid variables weren't visible out that far (and still weren't for Baade), so the brightest star in the Andromeda galaxy was used as a yardstick. When compared with the brightest star in a distant galaxy, it was assumed that those stars had the same actual brightness, so if they appeared (say) 1/16th of the brightness, they were four times the distance away.

The big advantage Sandage had over Hubble was 1950s photographic technology, capable of much clearer resolution. For many years, astronomers had not stared through lenses at the skies; they had made measurements on photographic plates, where it was possible to take a long exposure, increasing the amount of light collected from a distant star. Even though Sandage was mostly working on the same Mount Wilson telescope as Hubble, the photographic technology had moved on apace and produced much more detailed images. Sandage realized with what might have started as horror, but then probably turned into delight at his original discovery, that the "brightest stars" in many of those distant galaxies weren't stars at all.

Many galaxies contain huge clouds of hydrogen gas that haven't coalesced enough to form stars. However, over billions of years of sustained exposure to starlight, these clouds have gained energy, becoming hot enough to shine in their own right, although without the in-built nuclear reactions of a star to power that shine. It's a bit like heating up a piece of metal in a forge until it glows. The metal itself hasn't got a source of heat, but the energy it gains from outside makes it give off light. With the extra resolution he had available, Sandage could see that those "brightest stars" were often these huge glowing clouds rather than individual stars.

Once the clouds were eliminated, the brightest stars proved

much dimmer than first thought, and that meant each galaxy where this happened was farther away than Hubble and Baade had assumed. Although our near neighbor in Andromeda didn't shift, suddenly the gaps to the more distant galaxies were greater. The universe was now at least 5.5 billion years old, accommodating even the most recent age of the Earth. This became something of a lifetime quest for Sandage, who went on obtaining better data, which seemed to constantly push out the scale of the universe. By the end of the 1950s it was thought to be between 10 billion and 20 billion years old. For many years, although this estimate would be refined, the value would vary on a regular basis. It was only well into the twenty-first century that any confidence was given to a reasonably precise figure, based on a very different and much more theoretical method.

THIRTEEN BILLION YEARS

As recently as the turn of the twenty-first century, you could still see figures varying between 16 billion and 12 billion years, based on the radioactive decay of elements and the apparent age of the oldest observable stars, but the WMAP satellite (see page 148) is considered by many to pin this down with some certainty to 13.73 billion years, give or take just 0.12 billion years. However, it ought to be stressed that there is no magic indicator in the data that the WMAP satellite collected that can fix this date for certain. The dating relies on a model of the universe called the "Lambda–Cold Dark Matter model," which makes some significant assumptions about the nature of the universe.

This is a surprisingly simple model with only six parameters, numbers such as the Hubble constant describing the rate of expansion of the universe and the cosmological constant that shows the impact of dark energy, the mysterious force that is accelerating the expansion of the universe. Basing the measurement on this model does mean that if some of the other models of the universe we come across later are better than the Lambda–Cold Dark Matter model, the 13.7 billion years figure that you often see listed as fact is not based on anything substantive, and we would have to return to the rougher but better substantiated figures used earlier for the time span back to the Big Bang (or its equivalent in different models).

For some, however, arguing over the age of the universe was a waste of time. To them, the very idea that the universe began at a point in time was scientific heresy. They believed that it simply wasn't possible for the universe to emerge from nothing, particularly if this suggested a religious connotation, and as we have seen, it didn't help that Georges Lemaître, the originator of what would become the Big Bang theory, was a Catholic priest. When Hubble's work suggested the universe was expanding, an expansion that could be traced back to a starting point (see the next chapter), they insisted that this didn't fit with the universe as they understood it, so it had to be a misinterpretation of the data.

At this stage, there were still many holes to fill in when it came to the nature of the universe, but nowhere near as big as the uncertainty in how it came into being. After all, we are looking back billions of years. How can we be sure there ever was such a thing as the Big Bang?

5.

A BANG OR A WHIMPER?

The construction of the universe is certainly very
much easier to explain than is that of a plant.

—GEORG CHRISTOPH LICHTENBERG (1742–99)
Aphorisms (trans. R. J. Hollingdale)

Although the Big Bang is the most widely accepted scientific theory for the origin of the universe, there are some significant problems with the theory. It is a patched theory, one where the initial idea didn't match what was later observed, so extra components had to be thrown in to try to keep the theory working with what was known. And as with all cosmology—the study of the nature of the universe—it suffers from being based on circumstantial evidence, some of it so circumstantial that it would have little hope of standing up even in a court of law.

A SPECULATIVE SCIENCE

Cosmology suffers in comparison to most science because it is not experimental and there is no way to ever make it experimental. You can't do an experiment in controlled conditions on the universe as a whole and see what the outcome is. You have, instead, to make observations, often of something so far away in space and time that you can only ever study it highly indirectly, and then attempt to draw inferences. There is inevitably a fair amount of guesswork involved.

Rightly or wrongly, scientists add confusion to this picture by referring to the components of today's popular cosmological theories as if they were absolute truths. We can never truly prove something. I can't prove that the Sun will rise tomorrow morning; I can only say it is highly probable based on past evidence. The same goes for any scientific theory from Newton's laws of motion to Boyle's gas law. I can demonstrate that it matches all observations to date, but I can't prove it to be absolutely correct.

This is why it is so strange when those who dislike the theory of evolution complain that it is "just a theory." All science is just a collection of theories. Some might be given the weighty title of laws, but in truth they are all just best theories for the moment and likely to be modified or replaced in the future. This, of course, happened to Newton's laws, which were shown by relativity to be special cases that happen to apply when an object isn't traveling at close to the speed of light, which most of the time, for us, is the case.

So I can't prove anything in science, but I can easily disprove something. If I say that the Sun won't rise tomorrow morning and it does, then my theory is thoroughly squashed. If I find something doesn't obey Newton's laws, then under those conditions the theory doesn't apply. The only type of scientific proof with absolute certainty is when we disprove something. However, many scientific theories have a huge amount of experimental data to back them up, which makes them very useful. In cosmology, lacking those data, some of the predictions and suppositions may well be constructed on shaky ground.

For example, the existence of a list of things we tell about soon, including black holes, dark matter, dark energy, wormholes in space, and even the Big Bang itself, are all inferred. Theories exist that make it unnecessary for any of these things to exist. They may be out there; they may not. Cosmologists' most popular theories, and the data we have at the moment, suggest that they do exist, but it's a different level of certainty to the one we apply when we say, for example, that the Sun exists or that the Earth orbits it.

We sometimes hear about "the scientific method," as if there were only one way of doing science. Without experimental confirmation, cosmology has to take a subtly different approach. Given a hypothesis such as the Big Bang, scientists attempt to describe what the implications of such a hypothesis are, then look out for evidence that falls in with those implications. If what is observed differs, then either the observations are incorrect, the deductions that formed the implications were wrong, or the original hypothesis wasn't correct.

In the case of the Big Bang it might seem there's very little chance of finding anything measurable that could give us an insight into whether it really happened. After all, if it did happen, it took place billions of years ago, in a universe that bore no physical resemblance to our own, a tiny, unbelievably hot nugget of space, containing no matter as we now know it.

BIRTH OF THE BIG BANG

When cosmologist and priest Georges Lemaître first proposed the idea of the Big Bang it was from a logical tracing back of events. The Russian physicist Aleksandr Friedmann had shown from general relativity back in 1922 that space itself was likely to be expanding or contracting. With this basis, Lemaître was able to take the idea of an expanding universe as a possible starting point. Born in Charlesroi in the French-speaking south of Belgium in 1894, Lemaître originally intended to be an engineer, but his studies at the University of Louvin were interrupted by the First World War, during which he won the Croix de Guerre for bravery. When he returned to academia he switched to physics, with a particular interest in cosmology, and a few years later became a Catholic priest.

By 1927, Lemaître had picked up on Friedmann's largely ignored suggestion that space could be expanding. If that was the case, then earlier in its life, the universe would be smaller: earlier still, smaller still, and so on, until eventually it was a seed, a primeval cosmic egg of a universe. This idea also fit rather nicely with his Catholic beliefs, although there is no evidence that these had a direct influence on his formulation of the theory. But Lemaître found what he believed was corroboration of the theory in another recent discovery, the existence of cosmic rays.

Cosmic rays are showers of high-energy particles that spatter the solar system from out in the depths of the universe. If it weren't for the Sun we would be in serious danger from these particles, but the Sun's magnetic influence largely protects us from their impact, deflecting many of the particles out beyond our orbit. More still are mopped up by the atmosphere. Cosmic rays were discovered around fifteen years before Lemaître proposed his "primeval atom" theory, detected by a high-altitude balloon, which was exposed to more cosmic rays than reach the Earth's surface, as up in the stratosphere there is less protecting atmosphere.

It seemed reasonable to Lemaître that after the Big Bang, which he imagined to be the breakdown of a vast atom containing all the matter that was now in the universe, just as when an atomic nucleus fragments in a reactor today, there would be particles and energy thrown in all directions. Some of the matter, he assumed, would have been pulled together by gravity to form stars and planets. But the higher-energy particles would have resisted the relatively weak gravitational pull and would still be flying around as the aftermath of the great beginning.

SOLVAY PROBLEMS

Lemaître at thirty-three was not a youngster but was still relatively junior in academic circles. He was thrilled by the rightness of feel that his theory had. Scientists refer to the "elegance" of a theory, a notion that it just feels right, and for Lemaître this appeared to be an elegant theory for the origins of the universe. Not only did it emerge as a possibility from Friedmann's solutions of Einstein's general relativity, it now seemed to be backed up by the existence of cosmic rays. He had the opportunity to put his idea to Einstein himself at the fifth Solvay Conference in 1927. These were large-scale get-togethers of the big names in science, set up by the Belgian industrialist Ernest Solvay. Solvay originally intended them as a forum for the great and the good to discuss his own eccentric ideas on science, but Solvay was quickly, if gently, sidelined by the scientific heavy-hitters who attended.

The 1927 conference is largely remembered for being the meeting at which quantum mechanics really took off. Remarkably, more than half the twenty-nine attendees would win Nobel Prizes for their original thinking. These were big thinkers. It would seem this was a forum at which Lemaître would find a real opportunity for his theory to gain ground. Unfortunately, however, Einstein's mind was largely taken up with the challenge of attacking Danish physicist Niels Bohr's ideas on the new quantum theory, a theory that Einstein detested. In what little time he had, Einstein concluded that he did not like Lemaître's speculations. As we show, Einstein was so unhappy with the concept of an expanding universe that he tried to fudge the results of general relativity to iron it out. Yet here was this priest with a theory that depended on expansion.

This wasn't the only problem. If you took Friedmann's original work on the expanding universe, at the very beginning of things, the moment of the Big Bang, the universe would have occupied zero volume, sending the density of energy and temperature soaring up to infinity. It's not that uncommon for infinities to creep into physical calculations, but when this happens the assumption is usually that something has gone wrong, or at the very best, that science as we know it has broken down, and a different scientific regime applies. This meant that the Big Bang theory, taken to the logical limit, would have an insoluble enigma at its beginning.

Not only did Einstein come down firmly against Lemaître's ideas ("correct calculations" he remarked, but "abominable physics") he would ensure that the idea did not get far in the scientific establishment of the day. For Lemaître it was a crushing blow, and although he certainly didn't drop the idea, he would not try to get it further support until surprising observations were made by Edwin Hubble in the United States.

THE COLOR OF SPEED

Hubble was to crown his demonstration described in the previous chapter, that most nebulae were not gas clouds in the Milky Way but other distant galaxies, by the remarkable discovery that the farther away a galaxy was from us, the faster it was moving away. The apparently stately and fixed vista of the night sky was in fact blasting apart at phenomenal speeds, and this would prove essential to help Lemaître recover his theory and bring the idea of the Big Bang back into prominence.

The technology that made this observation possible relied on colors and was first used to discover with remarkable accuracy just what elements were present in stars, and then to get a feel for the way that stars and galaxies are moving with respect to us.

This ability to peer far out into space and discover what a star is made up of was the result of an effect of quantum physics, but one that depended on some discoveries made way back in the seventeenth century by British scientific genius Isaac Newton. Born in the Lincolnshire village of Woolsthorpe in 1642, Newton had a troubled upbringing before his mother grudgingly allowed him to go to Cambridge University as a sizar, a position that required him to act as a servant to a more wealthy student to earn his living.

Initially, Newton did not seem much of a student, but during a two-year break when the plague hit Cambridge in the winter of 1664, Newton made remarkable steps forward in ideas on light, math, mechanics, gravitation, and more. His work on light was inspired by a toy he bought at a fair. The Stourbridge Fair was held on common land down the river from Cambridge. It was just outside the jurisdiction of the university police, the proctors, which meant that students could have the good time they were denied in the city itself. Here Newton bought a crude prism.

In playing with that prism in a darkened room, Newton was the first to show how white light was made up of a mix of the colors in a rainbow spectrum, and it was also Newton who realized from this why a particular object looks a certain color. When we see, for example, a bright red fire hydrant, the sun's white light is hitting it. The hydrant absorbs most of the colors in the spectrum, leaving just red to be re-emitted. So we see the hydrant as red.

What Newton didn't know was why this was happening. We now know that the incoming light is a mix of photons with different energies, from the relatively low-energy red through to the high-energy blue. When a photon encounters the electrons that surround the atoms of matter (in the example above, the red paint on the hydrant) an electron absorbs the energy of the photon and jumps up to a higher level. Most of that energy will be gradually dissipated as heat in the object, but some of it will be used to produce new photons and these will have a characteristic energy associated with the material, in the case of the fire hydrant paint, the energy that gives a red coloration.

So when we shine light on a colored object, what it effectively does is to chop out a slice or a number of slices of the light spectrum and send those back out. But being lit up isn't the only way to see something. It can also, like a star, glow of its own accord. When this happens, the temperature of the object is sufficiently high that electrons are being forced to jump to higher levels by the energy from the heat. Sometimes they will drop back down, and the energy released will determine the color associated with the photon that is given off.

In a star, the temperature is such that there is a wide range of energies in the photons produced. But on the way out of the star, those photons pass through the star's outer layers. As they do, some of the frequencies are absorbed, and the result is a series of black lines on the color spectrum. (A little care has to be applied as other lines are produced as the light passes through the Earth's atmosphere.) Each element has its own characteristic black lines, and from these it's possible to deduce the elements that make up the star. These lines are detected using a spectroscope, a device which at its simplest is just a prism like Newton's, splitting apart the different colors in the light, and a microscope to examine the divisions in more detail.

Spectroscopy was first used to analyze the components of a star's outer layers, but spectroscopes would come into play in a different way when Hubble made his second great discovery. Here, the instruments would not be used to identify chemical content, but to track a shift in the color of the light. It's worth taking a moment to understand what's happening with the optical shift we see when a distant galaxy is moving. As we have already heard, this is related to the well-known Doppler effect. When a train passes at a railroad crossing, the train's whistle shifts downward in frequency, making a characteristic swooping noise. We hear a high-pitched note as it comes toward us, and a lower note as it moves away. The same is true of sirens on police cars and ambulances.

Something very similar happens with light. As an object moves toward us, the frequency of the light it emits goes up. You can envisage this happening by imagining a light wave flowing out from the object. Before the next ripple can come out, the object will be a bit closer than it was a moment ago, so the effect is that the wave is squashed up (that means a shorter wavelength and higher frequency); it is moved toward the blue. It undergoes a blue shift.

If you prefer, as I do, the photon view of light, then a blue shift is just an increase in the energy of the photons. The movement of the emitting body toward us gives the photons a boost of energy, just as a baseball thrown toward us from a moving car hits us with more energy than one thrown by a stationary pitcher. In the case of the baseball, that extra energy goes into extra speed, but light can't speed up. It can only travel at a single speed, the speed of light. Instead each photon gains a higher energy, a shift up the energy spectrum.

This means visible light emitted by something moving toward us is shifted toward the blue end of the spectrum: the light gets bluer or more energetic. As an object moves away from us, the light it gives off is shifted toward the red, lower-frequency end of the spectrum. It has less energy. If it weren't for spectroscopes, it would be impossible to measure such a blue or red shift. If, for instance, you see a red star, how do you know it's not just a star that happens to be red, rather than one that is red-shifted?

The answer is down to those patterns of black lines in the spectrum of emitted light. The pattern of these lines is like a fingerprint. The expected pattern for each element is well known, and can be identified even if the light is shifted in color by a blue or red shift. So we can look at the light from a star or galaxy and see how much it has been shifted. From this, it is not difficult to work out how fast the distant body is moving with respect to us.

This technique wasn't new when Hubble made use of it. The British astronomer William Huggins, who with his wife Mary first realized how much could be discovered about stars as well as the Sun, where spectroscopy was first used in astronomy, was also the first to realize that any shift in position of those definitive lines could be used to identify the relative speed of a light-producing body out in space when in 1868 he detected a red shift in the star Sirius.

HUBBLE'S LAW

We still haven't quite reached Hubble's discovery. With his new confidence in the scale of the universe, Hubble was prepared to take on some data that had been pulled together by the Indiana-born astronomer Vesto Slipher before the First World War. Slipher had measured large red shifts in a good range of nebulae, and large blue shifts in a few more. His figures, which suggested that these objects that were thought to be part of our own galaxy were hurtling all over the place at incredible speeds, seemed impossible to interpret effectively. It took one last push from Hubble to suddenly open up the prospect on a universe that was not just immense but expanding.

With his assistant Milton Humason, as different a man from himself as was possible to imagine, Hubble set out to unravel the mystery of the high-speed galaxies. Where Hubble's success in determining the scale of the universe, and his roaring self-confidence, had made him a celebrity scientist to rival Einstein, Minnesota-born Humason had dropped out of school at age fourteen, and had started off at Mount Wilson as a mule driver, carrying provisions and equipment up the mountain to the observatory. Fascinated by what he saw going on in this exotic location, Humason managed to absorb what he could by hanging around the observatory and over time began to be trusted with more and more observations. Between them, with Hubble's dramatic style and English airs and graces, and Humason's dogged determination and American true grit, they were ready to take on the universe.

The result of many, many nights of observation and calculation was shattering. Not only were most of the other galaxies moving away from ours, the farther away they were, the faster they receded. And there was more still than that. Within the limits of the error of their measurement, there was a neat linear relationship between distance and speed. Double the distance away and you doubled the rate at which the galaxy was racing away from us. This beautiful, surprisingly simple relationship became known as Hubble's law.

EXPANDING EVERYWHERE

As we look outward, the vast majority of galaxies are red-shifted. They are moving away from us, and the farther out you get, the faster they are moving. This doesn't imply some kind of return to the special position the Earth was awarded in ancient times as the center of the universe. As it's the space that makes up the universe that's expanding, not the galaxies moving away from us within space, every part is expanding away from every other part. That probably needs a little clarification.

The simplest way to think about this is to imagine a two-dimensional equivalent. Imagine a flat sheet of rubber with sequins stuck to it. Get someone holding each corner of the sheet, and begin to stretch the rubber in all directions at once. Then, whichever sequin you pick to represent Earth, every other sequin is moving away from you. (This is sometimes described as being like the surface of a balloon expanding, but I think this is confusing as the expansion in both examples is in the two dimensions of the rubber, not an expansion of three-dimensional space as we see in the universe.) Note also that the sequins aren't moving with respect to space (the rubber): it's space itself that expands. A sequin doesn't move when compared to the piece of rubber underneath it.

Hubble still found that a handful of galaxies, such as the M31 galaxy in Andromeda, weren't moving away, but were heading toward us. As Slipher had suggested, they were blue-shifted. These proved to be the very nearest galaxies, ones where the expansion of the universe had least effect, and so could be overcome by the force of gravity. The universe is still expanding between us and them, but these galaxies are heading toward us through space faster than that stretching of space can send them away.

For Hubble, the implications of his discovery weren't hugely significant. He took his lead from that greatest of English scientists, Isaac Newton, who was inevitably something of a hero for Hubble. In his masterpiece the *Principia Mathematica*, published in 1688, Newton had said, speaking of gravity, "*hypotheses non fingo*," meaning, "I frame no hypothesis." (Frame, incidentally, was a word with negative connotations then; he was saying, "I don't fake explanations.") Newton was saying that he made measurements and provided a description of how gravity worked, but did not guess at what mechanism made it work that way. Similarly Hubble was happy to have established the relationship in speed between galaxies, but was not prepared to follow this through and suggest that this meant everything had started in a Big Bang.

However, if Hubble's view of the universe as we see it now was accurate, it was only reasonable to suggest that the cosmos started from a single location in time and space, a moment of universal creation, the Big Bang. Follow back each of those galaxies and the trail led to the same place—just as the trail from our own galaxy does—a single location for the beginning of everything.

This is why, incidentally, we can't point to a place in the universe and say, "That's the center of everything. That is where the Big Bang happened and everything expanded out from there." Because the whole of space has expanded out from that point, we are all at the point where the Big Bang happened; what is now all of space corresponds to the point where the Big Bang happened. It has just stretched out in all directions like three-dimensional taffy.

EDDINGTON'S U-TURN

The possibility that there really was a beginning from a single location wasn't initially a popular one. Lemaître's idea of a primeval atom had been thoroughly squashed once Einstein gave it the thumbs-down. Although it might seem obvious now that the expansion of the universe implies some kind of beginning, it was not a common interpretation of Hubble's findings at first.

For the moment, just as they had with Slipher's original discovery of galactic red shifts, the astronomical community was happy to leave this as a mystery. But for Lemaître this was a vindication of his theory and he looked for support in reestablishing the image of an explosive start to the universe from a single point. In his early days, it had been the great British astronomer Arthur Eddington who had offered support for his ideas of an expanding universe, and it was to Eddington that he returned now that he had Hubble's data to support his primeval superatom theory.

A great scientist needs many skills and traits, but one of the most important is the ability to change direction at a moment's notice. This may well be the reason that most politicians are so bad at understanding and supporting science. In politics, making a change of direction is considered a display of weakness. If politicians change policy midstream, they are attacked because they clearly didn't know what they were doing the first time around. The reverse is true in science, where failing to change direction when the evidence warrants it is seen as weakness. The best scientists find it easy to accept that they were wrong and move on. A theory (any theory, however much is invested in it) is only as good as the data that support it.

Eddington showed exactly the trait of the good scientist in his reaction to Lemaître's pulling together of his primeval atom theory and Hubble's expansion evidence. Together they made Eddington's original dismissal of the theory out of date. Eddington was big enough to admit that not only had he dismissed Lemaître's theory at the time, but he had forgotten all about it, and so didn't spot the relevance of Hubble's new data until Lemaître wrote to him again.

Now, though, Eddington demonstrated the great scientist's ability not only to change direction, but to throw all his weight and enthusiasm behind a new theory. He wrote to the leading scientific journal, *Nature*, pointing out the significance of this new information and the weight it lent to Lemaître's theory, and translated Lemaître's paper from the French to republish it in a British astronomical journal.

When we now look back at Eddington he seems a stuffy, archetypal old-style Englishman, but his widespread popularity at the time was largely down to an uncommon ability in a scientist to be able to explain scientific matters in terms the layman could understand. He was good at coming up with the right turn of phrase. It was Eddington who, according to legend, is said to have showed his grasp of communication skills when asked by a journalist if it was true that he was one of only three people who understood Einstein's general relativity. Eddington did not reply, and the interviewer pointed out there was no need to be modest. Eddington corrected him instantly. He wasn't being modest; he was trying to think who the third person was.

Apocryphal or not, this story marked the sentiment of the time on Eddington's knowledge, although some seemed to doubt whether even Eddington could comprehend relativity. When he made a presentation on general relativity to the Royal Astronomical Society in December 1919, physicist Oliver Lodge responded, "One of the things which astonished him was that Professor Eddington thought that he understood it." *The Times* newspaper notes that this got a laugh from the audience.

Eddington's sense of tongue-in-cheek fun also seems to come through in his impossibly precise remark years later when giving a lecture: "I believe that there are 15,747,724,136,275,002,577,605, 653,961,181,555,468,044,717,914,527,116,709,366,231,025,076,185, 631,031,296 protons in the universe, and the same number of electrons." However, in his later years, Eddington did focus strongly on arguments that led him mathematically to this remarkably precise figure, so he might not have conceived it as a humorous statement.

When it came to the Big Bang and the expansion of the universe, it was Eddington who first gave a picture to understand this expansion that was approachable by the general public, the idea of dots on the surface of an expanding balloon, rather like my sequins on a sheet of rubber.

As we have seen, the downside of Eddington's picture is that our universe isn't a two-dimensional object distorted in three-dimensional space as is the balloon. The surface of the balloon has no depth; it just extends in two dimensions, but we have distorted it in a third to make it contain a sphere, and it's in three dimensions that we blow it up. Unlike the balloon, the universe starts off in three dimensions, so it is harder to envisage that expansion taking place, but it still is space that is expanding, not the galaxies moving within space (except where gravitational attraction confuses the results, such as the attraction between the Milky Way and the galaxy in Andromeda).

MOVING FASTER THAN LIGHT

A strange effect falls out of the way that this expansion of space takes place. It can produce an apparent contradiction of Einstein's special relativity. Special relativity says that nothing can move faster than light. However, expansion of space makes it possible for light—or physical objects—to exceed the light-speed barrier. The limitation of light's speed, around 300,000 kilometers per second, is within space. If space itself expands, objects can get to move faster than the speed of light with respect to each other as a result of the distortion of space itself. Within space they might not be moving at all, just like the sequins on the rubber sheet.

A rather similar effect can be seen when crossing the Atlantic eastbound on a plane. Because of the high-speed winds that form the jet stream, the plane can be traveling significantly faster compared to the ground than it is relative to the air around it. A conventional aircraft can't travel faster than the speed of sound without severe damage, but I have flown on a 747 that flew faster than sound with respect to the ground, even though its speed with respect to the air around it was only the usual cruising speed. Similarly, the galaxies in the expanding universe never exceed light speed with respect to space, but might travel faster than light with respect to each other.

REASONS TO RED-SHIFT

Now the bandwagon was growing. A year later, Einstein, who had originally squashed Lemaître's ideas, jumped on board, supporting the expanding universe and accepting, just as Eddington had, that he had previously been wrong in insisting that the very spacetime of the universe was static and unchanging. However, it shouldn't be thought that Lemaître had won everyone over. When he and Einstein came together in a seminar near Mount Wilson in 1933, there were still a good number of people who stuck to the idea of that expansion-free universe, and more still for whom the idea of a dramatic explosive beginning to everything simply didn't fit with their picture of reality.

This was fine. Science happily accepts alternative theories, but to have a chance of existence, those theories have to explain the observed data. One substantiated bad datum can be enough to blow a scientific theory out of the water. Those who argued that the Big Bang idea was wrong and still believed we lived in a static universe had to explain why, in such circumstances, the galaxies were red-shifted.

The most ingenious suggestion came from firebrand cosmologist Fritz Zwicky. Born in Bulgaria of Swiss parents in 1898, Zwicky worked at the prestigious Federal Institute of Technology in Zurich, Switzerland, before emigrating to the United States in 1925 to join Caltech and the Mount Wilson Observatory. Zwicky's picture of what was happening is better understood by taking the photon view of light, where a red shift is a reduction in the energy of photons.

He argued that the gravitational pull that the photons heading for us felt from the rest of the universe reduced their energy and red-shifted them. The farther away they were, the more opportunity there was for the gravitational influence to build up, so the bigger the shift. It's certainly true that gravity will red-shift light for exactly this reason, but there was no explanation of where such a huge energy loss was coming from. Gravity is a very weak force compared with the other forces of nature.

It was here that Zwicky crossed from good science to bad. As we've seen, a great scientist is one who can turn on a dime, changing direction from one theory to another when the data require it. Zwicky instead hung on to his theory and challenged the data that disagreed with it. Putting up a challenge to the accuracy of data is perfectly acceptable; that's why science requires experiments to be replicable, because otherwise you could be relying on spurious numbers. But in this case, for Hubble to have been wrong enough to make Zwicky's theory work, he would have had to have made massive mistakes with his measurements. Not only was there no evidence of this, there was increasing corroboration of Hubble's results.

When a theory is first proposed there is often insufficient support to make it worth testing out the consequences, but once it gains the sort of acknowledgment the Big Bang had, scientists can begin to test it against the various observations already made about the universe to see whether the theory is supported by the observation. One immediately sprang to mind.

There had always been a slight sense of unease about the odd distribution of the chemical elements in the universe. There is vastly more hydrogen and helium in the universe than there is of any other element: only around 0.1 percent of all atoms in existence make up everything other than hydrogen and helium, including everything we consider part of the solid visible world we inhabit. The prevailing idea of a fixed unchanging universe offered no explanation as to why these particularly high quantities of hydrogen and helium should occur.

There had to be some value, of course, but it would help to support a theory if it predicted that the observed values should be true. Once the Big Bang was taken seriously, even if it were not yet generally accepted, it seemed sensible to see if it could offer an explanation for this strange universal composition of matter through the universe. The man who would provide an answer was George Gamow.

THE HYDROGEN SEA

Gamow was one of science's most ebullient characters. Born in Odessa in Ukraine in 1904, he began his work in physics at the local Novorossia University, but was soon frustrated by the way the Soviet system selected scientific theories as true more because they were politically acceptable than because there was evidence for them. He made two failed attempts to get out of the Soviet Union (in one trying to paddle a canoe more than 150 miles across the Black Sea to Turkey with his wife) before managing to defect to the United States at one of the Solvay Conferences.

Freed of the shackles that the old Soviet system imposed on scientific free thinking, he would prove, like Eddington, to be one of the few great scientists who was also a good popularizer. Gamow wrote a series of books about a fictional Mr. Tompkins who experiences dream worlds where the laws of physics are different from our own, using this setting to explain much of fundamental physics to young audiences. These books, and another he wrote on the chemistry of life, would prove an inspiration to a whole generation of younger scientists.

It was Gamow who infamously persuaded another famous physicist called Hans Bethe to reluctantly add his name to a paper Gamow published with Ralph Alpher, purely so the paper could have the list of authors Alpher, Bethe, and Gamow, sounding so like the first three letters of the Greek alphabet, alpha, beta, and gamma. Alpher, who was much less famous than the other two, was horrified by this stunt, as he was bound to be sidelined by two big names, one of whom hadn't even contributed to the research. I am afraid that I also sideline Alpher to the extent that I often refer to their joint work as Gamow's from now on, but please don't forget poor Ralph Alpher. It was this paper that attempted to show where the abundance of hydrogen and helium in the universe originated.

Even in their explanation, Gamow and Alpher made one big assumption: that the universe started off as a sea of hydrogen. This was almost the diametric opposite of Lemaître's primeval atom. In Lemaître's picture of the very beginning, there was an immense atom containing all the nuclear particles that would make up every future atom. The Gamow picture started with the simplest building blocks there were, a sea of the most basic atoms, hydrogen.

This is nice and simple, appealing to the scientist's sense that a good solution is an elegant solution, but that doesn't make it right. Gamow had not explained where that hydrogen came from. He merely took the question mark one stage back, from a universe composed of hydrogen and helium to one that was pure hydrogen.

MAKING HELIUM

Gamow realized that if things had started as just hydrogen, then the first suspect for the production of the helium that is also present in large quantities had to be all those billions of stars out there. It was already known that most are in the business of churning hydrogen into helium. Our Sun alone makes half a billion tons of helium a second. Yet the output of all the stars just wasn't enough. If the Sun were to produce all the helium it already contains this way, it would have to burn for twice what's currently thought of as the lifetime of the universe (a lifetime of over 13 billion years), and the universe is now thought to be ten times older than it was when Gamow was working on this topic.

Instead, then, Gamow imagined that most of the universe's helium came out of the Big Bang itself. (Helium was given its name because it was first discovered in the Sun, thus perhaps it should therefore more accurately be renamed bangium.) He imagined that the temperature was such at that early stage of the universe's life that the sort of processes we see in the Sun—nuclear fusion—took place. Fusion is an entirely different form of nuclear power to the fission process that drives existing nuclear power stations. In fission plants, the nuclei of atoms are blasted apart by high-energy particles, generating both heat (that will be used to generate electricity) and further high-energy particles that will cause more nuclear fission in a chain reaction. Fusion is a different process.

All the Sun's majestic power comes from the conversion of hydrogen to helium using nuclear fusion, where the merging of particles to provide a new element results in the release of energy. In order to fuse, the positively charged protons that make up the nucleus of hydrogen have to be squeezed incredibly close together. But even in the pressure and temperature present in the Sun, there isn't enough energy to overcome the repulsive force that keeps protons apart. They are positively charged, and as with bringing two identical magnetic poles together, the result is that they repel each other.

It is only because of one of the oddities of quantum physics that stars work at all. Quantum particles like protons don't have an exact location unless they are directly observed. Each particle is spread out over a range of locations, with a different probability of being in any one of those locations. This means they can "tunnel," jump from one place to another, even if there is a barrier in the way, *without passing through that intervening barrier.*

This is what happens in the Sun. It is only because protons can tunnel through the distance that their repulsion keeps them apart, instantly emerging so close to another proton that they can fuse before they bounce away from each other, that the Sun can shine.

Such tunneling has a low probability of occurring, but there are so many protons in the Sun that it's happening all the time on a huge scale. When nuclei fuse together the result is a small loss of mass which is converted into energy. This is a much cleaner approach to nuclear power, which would be far safer and less polluting than the current fission plants, but successive governments have failed to invest enough to make fusion plants a practical possibility. In a star, though, or in the early universe in the Big Bang model, there is no need for funding: it just happens.

Gamow imagined that in the first few moments of existence after the Big Bang, hydrogen ions (atoms that have lost electrons) would fuse together in the immense heat and pressure to form helium, and that this was the source of much of the helium in the universe. He used the same kind of reversed-time calculation that was used to decide how old the universe was, to work out what the conditions in that early universe were like.

Imagine running the universe backward. Instead of expanding, everything gets squashed closer and closer together. With compression comes an increase in temperature, which is just a measure of the energy of the atoms involved. Go far enough back and the pressure and temperature would be so high that the same kind of fusion as occurs in stars could take place. The whole universe would be like one immense star. (I've referred to Gamow for compactness, but it was Alpher, his junior, who handled the heavy-duty math while Gamow concentrated on the big ideas.)

This is still thought to be the case, although Gamow's greater hope had been to show that all the different elements came into being from the Big Bang, and this proved untenable. Conditions weren't severe enough in that initial primordial soup: it takes the immense temperature and pressure of a collapsing star to cook up the heavy elements, which were then blasted out into space. The heaviest elements, such as uranium, were formed only as stars exploded into supernovae. Without these ancient stars, predating the Sun, there would be no Earth, no oxygen to breathe, no carbon to build living molecules, and no silicon to support our modern electronic world. All the Big Bang had a hope of producing from that sea of hydrogen were three other elements: helium, lithium, and beryllium.

BIG BANG AND BLACK BODIES

The abundance of helium wasn't the only prediction that could be tested against the Big Bang theory to see how it held up. Gamow had another idea up his sleeve. To be precise, Ralph Alpher had an idea in conjunction with another physicist, Robert Herman. But Alpher had, as he suspected, remained largely off the radar thanks to Bethe's inclusion on the original paper with Gamow, so it wasn't until Alpher and Herman got Gamow on board that their theory could truly be tested on the scientific community.

The trio had a hunch that from the beginning the Big Bang would give off what is known as blackbody radiation. This radiation sounds much grander than it is, describing how an object absorbs and gives off light. To see what it's all about, we need to return briefly to why different objects appear to be differently colored. Why, for example, is a school bus yellow? You might say it's because it's painted yellow, but that's not what I mean. What do we mean when we say it's yellow?

As we've already seen, Isaac Newton, that most versatile of early scientists, was the man who originally answered this question. He had already worked out that ordinary white light is much more complex than it seems. When white light passes through a prism (or the raindrops that form a rainbow) we get a whole spectrum of different colors, each and every one of which is present in that white light. So imagine all those different rainbow shades in the light hitting our school bus. Let's zoom in on the atoms of paint on the outside.

What Newton didn't know, but the great American scientist Richard Feynman would explain, is that when light hits an atom the tiny packets of energy in the light (photons) are absorbed by the atom, giving the electrons on the outside of the atom extra energy. But atoms are fussy and accept energy only in certain amounts. The photons that have the wrong energy are thrown out. And the energy of photons comes across as color. The different colors we see are caused by photons of different energy.

As it happens, the paint on the school bus is fine at absorbing photons with the right energy to come back red or blue or green but it's not interested in yellow photons. (In fact it can be a bit more complicated than this, as different mixes of photons are perceived by our eyes as different colors, but let's keep it simple.) The photons that aren't absorbed are the ones that trigger our eyes to see yellow. So we say the bus is yellow.

Now imagine that instead we had a bus that absorbed every single photon that hit it. Nothing comes back to our eyes. Then this bus would be in physics terms a blackbody. This is more than just being painted black. However good black paint is, a few photons will come back out, but we're saying every single photon that hits our blackbody bus disappears. It would be truly black, a void, and an emptiness of sight.

RADIATING BLACK

So that's a blackbody, but how about blackbody *radiation*? We're not talking here about radiation in the frightening atomic sense;

this radiation is light. Visible light is just a small part of the huge spectrum of electromagnetic radiation—photons with different energy—that ranges from low-energy radio all the way up to high-energy X-rays and gamma rays. The "radiation" in black-body radiation is just light. But radiation from a blackbody seems not to make any sense. We've already said that a blackbody absorbs every bit of light that hits it and doesn't let it out, so how can it radiate light?

This is because there are two separate reasons that matter appears to have a color. One is how it treats those incoming photons. But the other way it can have a color is by generating photons all of its own. As you heat up an atom you push energy into its electrons, and every now and then, one of those energized electrons will pump out a little packet of that energy in the form of a photon. The object begins to glow. The hotter the object, the more energy the photons have. So as something gets hotter and hotter it glows red, then yellow, then white. Blackbody radiation is the pure output of this kind of glow. We start with something that doesn't let out any light that hits it, so there's no confusion from reflected light, and we just see the photons that are pushed out by the heat in the body.

So imagine an incredibly hot Big Bang. Initially, once things had cooled enough for atoms to form, the universe would be a sea of plasma. Plasma is one of the states of matter. One quick potential for confusion needs clearing up here. This has nothing to do with blood plasma. (Actually neither of the uses of the word works particularly well with its origin, as originally "plasma" meant something formed or molded, and both types of plasma are very obviously lacking form.) Blood plasma is the colorless liquid in which blood corpuscles float; it's the liquid part of blood. Plasma in the physics sense is the fourth state of matter, a more energetic form of matter than a gas.

To show how poorly plasma is understood, my dictionary defines plasma as being a gas in which there are ions rather than atoms or molecules. Let's not worry for a moment about those ions, but note how the dictionary was thinking loosely. To give such a definition is like calling a liquid "a very dense gas with fluid properties." A plasma is more like a gas than it is like a liquid, just as a gas is more like a liquid than a solid, but it is still something else, a different state of matter.

In practice we tend to have more direct visual experience of plasmas than we do of gases. The sun is a huge ball of plasma. Every humble candle flame contains some plasma, although they're pretty cool in plasma terms, so flames are usually a mix of plasma and gas. Just as a gas is what happens to a liquid if you continue to heat it past a certain point, so a plasma is what happens to a gas if you continue to heat it far enough.

As the gas gets hotter and hotter, the electrons around the atoms in the gas are bumped up to higher and higher energy states. Eventually some have enough energy to fly off as independent particles. In general, depending on how many electrons they have farthest away from the nucleus, atoms have a tendency to find it easier either to lose one or more electrons or gain one or more electrons. Atoms that easily lose electrons do so, and end up as a positively charged ion. Atoms that easily gain electrons suck up the spare electrons from the positive ions and end up negatively charged. This is a plasma.

Plasmas are very common once you consider the universe as a whole. After all, stars are pretty big objects. Up to 99 percent of the universe's detectable matter is plasma. Although plasmas are gaslike in not being hugely dense, they are very different from gases. For instance, gases are fairly good insulators; plasmas are superb conductors.

When we look back in time to shortly after the Big Bang, the universe was filled with plasma. As we've seen, plasmas (unlike gases) are good conductors. And good conductors tend to be opaque because materials with all those free electrons floating around in an ion soup are great at interacting with photons of light. This is also true of plasmas: they scatter light.

BECOMING TRANSPARENT

This term "scattering" gives the impression that light is bouncing off the electron. In fact photons don't bounce off in the scattering process, any more than light bounces off a mirror. What actually happens is that a scattering electron (just as one of the electrons in the mirror surface) absorbs the energy of the photon that hits it. The photon is destroyed. Because the electron has gained energy it makes what is known as a quantum leap, jumping from one level of energy to another.

However, electrons in high energy levels tend to be unstable. Very quickly some or all of the energy is spurted out, dropping back to a lower level. A new photon comes shooting out in a different direction. The light has been scattered. If the electrons are tied up with the nuclei in the plasma, it is harder to get them to change energy levels. The photons are much less likely to be scattered. So as the post–Big Bang plasma cooled to form a gas, the electrons would no longer scatter photons; the universe would become transparent.

It was only as the post–Big Bang plasma cooled sufficiently for complete atoms to form that it became transparent, and all the light energy generated in the beginning was allowed free. Alpher and Herman calculated that this happened around 300,000 years after the Big Bang. At the time that plasma changed to gas, they estimated the temperature would be around 3,000°C (5,400°F). This would make the light from its blackbody radiation just outside the visible range in the near infrared.

COOLING LIGHT

Over billions of years, the apparent temperature of this light would drop. At first sight this is a bizarre concept: light doesn't have a temperature, and it doesn't "cool down" as it travels. One of the things that makes light special is that it can cross millions of light-years unchanged. It only loses energy when it interacts with something, and then it's all or nothing. But as we have seen (page 89) the expansion of the universe would have the effect of red-shifting the light, extending its wavelength or dropping its energy. The apparent temperature, which is just a measure of the energy of a material, would fall because of that loss of energy.

Gamow, Alpher, and Herman made a calculation of the kind of energy they expected the photons left over from the origins of the universe to have now. The red shift put them well below the visible, in the region of electromagnetic radiation we call microwave. That's the same stuff that heats up food in your microwave oven.

Temperature is just a measure of the energy in the body that's doing the radiating. The equivalent temperature of the microwaves Gamow expected to see was around 5 kelvin. Kelvin are units of temperature with the same size of degree as degrees Celsius (centigrade), but starting at around −273.16°C. This is absolute zero, the coldest possible temperature, which is the equivalent of zero thermal energy. The radiation Gamow predicted would be reaching us from the Big Bang would be just 5 degrees above absolute zero.

If this radiation could be detected it would add useful weight to the Big Bang theory. This radiation should, according to Gamow, be blackbody radiation, which has energy distributed in a characteristic way, and it should come from everywhere, because whichever direction you look, when you look far enough you should be looking back to the beginning. At the time, however, no one knew how to detect such energy and Gamow's idea was put to one side and forgotten.

THE UNIVERSAL SEED

Although Gamow's ideas could not be taken any further at the time, there was still plenty of opportunity to think about the origins of the universe and what the initial conditions in the Big Bang would be like. Exciting though the idea of everything flowing from a tiny cosmic egg is, it's not a theory that obviously fits what we see out in the universe. To the casual observer, the most obvious problem is that there is just too much universe out there to have ever fitted in a tiny "superatom." Although most of space is empty, you can't just compress everything closer and closer together.

We know that even with a small body (by universal standards) such as the Sun there is a limit to how much matter can be compressed together. That's why, before quantum tunneling was discovered, it was first thought that the Sun shouldn't be able to operate by nuclear fusion. Given we can't even compress all the matter in the Sun much more, how could cosmologists possibly imagine fitting all of the matter in the universe, thought to contain at least 100 billion galaxies, each containing on average maybe 100 billion stars, into a tiny speck?

Lemaître's original picture didn't have this problem, because his "primeval superatom" was a huge object, containing every single proton, neutron, and electron that now exists. It might have been an atom, but it was (relatively) enormous. As the bigger an atom is, the more unstable it is, it would have broken down very rapidly to smaller, more stable atoms. However this picture didn't fit with a universe that was mostly hydrogen and helium, as ours is. Once the superatom broke down, Lemaître's early universe should have been filled with heavy atoms. So an alternative view of the cosmic seed was developed that really did start with a tiny—in effect zero-sized—speck.

Before looking at the beginning according to the current Big Bang theory, though, it is worth briefly mentioning one alternative that was closer to Lemaître's original picture. This idea was developed by physicist Ernest Sternglass and is unknown to many modern cosmologists. Sternglass, it should be emphasized, was a serious physicist, not a crank, but his theory has never been adopted and has probably never been tested enough to be absolutely sure that it is incorrect.

Like Lemaître, Sternglass imagined starting from a single "primeval atom," but Sternglass was able to build on his more advanced knowledge of relativity and quantum theory to make this incredibly massive particle out of what would normally be two almost unnoticeable particles: an electron and its anti-matter equivalent, a positron.

ENGAGE THE ANTI-MATTER

Although anti-matter tends to be featured as a power source in science-fiction movies, it's a real enough concept. Anti-matter is the same as ordinary matter, but the particles that make it up have the opposite electrical charge to those in conventional matter.

Where, for example, an electron has a negative charge, the anti-matter equivalent, the anti-electron or positron, has a positive charge. There are similar, anti-matter equivalents of all the charged particles. When two opposite anti-matter particles (an electron and a positron, for example) are brought together they are attracted, smash into each other, and are destroyed, converting all their mass into energy. This doesn't happen when a negative electron meets a positive proton because there are nuclear forces in place to prevent annihilation, but no such force protects matter and anti-matter.

Sternglass imagined these two particles forming a sort of primitive atom, with the electron and the positron orbiting each other at very high relativistic velocities, producing a single particle with incredibly high mass. This particle would then split to produce two superparticles, still vastly massive; they would split again, and so on, rather as a single cell splits again and again to make a human fetus. Some of the early splits of the universal primal cell would be responsible for the galactic clusters and galaxies we see now; later ones would produce the apparently fundamental particles with which we are now familiar.

Sternglass even imagined some delayed mini–Big Bangs resulting from the splits that would produce vast bursts of gamma rays, a phenomenon that has been observed in distant time and space. This theory has the benefit over many others of being intensely simple. It's back to having really basic building blocks from which everything else eventually evolves. It should be stressed, however, that although it was simple it was not simplistic. This was not like a nonscientist thinking, "Hey, everything must have come from a simple pair of particles at the start." Sternglass based his ideas on good physics, and up to the mid-1990s when he last worked on it, the theory was consistent with both physical theory and observed cosmological data.

Sternglass' ideas have not been taken forward. They were probably a dead end. And he offered no explanation for where the high-energy electron/positron pair came from in the beginning. Yet I wanted to include them to show just how much our accepted ideas, whether it's Big Bang or stranger ones we meet later, are very much just the theories that have been followed up. We could still be waiting for *the* idea that will sweep all the rest away, or it could even be that Sternglass was right. Eventually, science will get there, but it won't necessarily be a quick and clear progression from where we are now. Real science is much more messy and undirected than science in the movies.

THERE IS SUCH A THING AS A FREE LUNCH

Back with the accepted Big Bang picture, to get around the problem of cramming everything into this tiny space, astrophysicists adopted a cunning plan. They wouldn't get everything in at all, but instead they would resort to making an assumption that might seem like cheating, but enabled them to suggest that the original universe had very little matter in it at all. They achieved this sleight of hand, apparently creating something from nothing, thanks to a strange way of looking at gravity: you can consider gravity to be a form of negative energy.

If that's possible (and we show how in a moment) then all the gravity out there in the universe can be used to offset much of the matter. If you consider matter to be nothing more than positive energy (and Einstein told us with $E = mc^2$ that matter and energy are interchangeable), then you can imagine matter coming into being effectively from nothing as gravity's negative energy balances it out.

If you aren't convinced that gravity is the equivalent of negative energy, consider pushing a satellite into orbit. We have to put energy into the system, generated by our rocket, to get the satellite off the earth and into a stable orbit where the gravitational pull is lower. So to move from one stable state to another we've put energy in, the pull of gravity we've fought against can be considered as negative energy that balances out the positive energy put into the system by the rocket, resulting in energy being conserved overall, a requirement that physics imposes on us.

In science classes at school you might have learned about potential energy. If I hold a weight over your head, that weight is said to have potential energy due to its position. If I then let go it will drop and hit your head. Its potential energy will have reduced, converting that energy into the kinetic energy of motion which will then be translated into painful mechanical energy as it hits your head. Saying the weight had positive potential energy is just a mirror way of saying that falling increased the negative energy due to gravity.

THE PROBLEMS WITH BIG BANG

Getting all of the matter in the universe from practically nothing isn't the only challenge faced by those who wanted to make the Big Bang theory work. There is also the problem that the whole universe is too uniform. The temperature across the universe doesn't vary a lot, and on the large scale, matter is fairly uniformly distributed across it. Yet to have uniformity, different parts of the universe would have to communicate: either to be in direct contact, or to get information from one place to another to produce this similarity. And the universe is so big, there hasn't been time in its lifetime for this smoothing out to occur, at least if the universe expanded in the past the way it is now.

Similarly, the universe is too flat. This may seem a strange concept for a three-dimensional space, but here we're talking about curved space in the sense Einstein dreamed up. According to Einstein, space is curved by adjacent matter, the way a sheet of rubber is curved when you put a bowling ball in it (only space is curved through three dimensions, rather than the two-dimensional sheet of rubber). It's this curvature that causes gravitational attraction as objects slide down the curve. However, if the whole universe started out as a tiny primeval superatom, the initial curvature would be immense, and again, extrapolating back, the current rate of expansion would leave space much more wrinkled and bumpy than is actually the case, with its near flatness.

Such was the concern about these issues that it was quite easy in the early days of Big Bang for an alternative theory to challenge it, and the three scientists who developed its first big opponent would be influenced by a night at the movies.

6.

KEEPING THINGS STEADY

This [photograph of a very large flock of geese] is our view
of the conformist approach to the standard (hot big bang) cosmology.
We have resisted the temptation to name some of the leading geese.

—FRED HOYLE, GEOFFREY BURBIDGE and JAYANT V. NARLIKAR,
A Different Approach to Cosmology

In the early days of the Big Bang theory, an alternative was posed by three astrophysicists usually fronted by Fred Hoyle, the bluff British scientist who later explained how the heavier atoms are made in stars and supernovae and who came up with the name "Big Bang." Hoyle was an unlikely academic for the time. Born in Gilstead, a village on the moors above the undistinguished town of Bingley, he had many of the characteristics that are stereotypically applied to Yorkshiremen.

Sometimes called the Texas of the United Kingdom, Yorkshire is the biggest English county, and has a fierce sense of independence, made all the stronger by being the losing side in the Wars of the Roses, defeated by its archrival Lancashire. Hoyle was born in 1915 and like a number of other leading scientists proved rebellious in his early attitude to education, spending more time learning from the world than in the schoolroom. However, his interest in astronomy, sparked by a friend of his father's who owned a telescope, seems to have made him realize he had to settle down if he were to achieve anything. By the time he was high school age, Hoyle was back in an education system that would take him to the country's top university for science, Cambridge.

Apart from a wartime excursion into radar research, Hoyle would remain at Cambridge for many years. He was anything but a typical occupant of academia. Like Richard Feynman in the United States, he had a rough independence of spirit that didn't fit well with an Ivy League environment. Where most Cambridge academics of the time would have had cut-glass accents and starchy manners to match, Hoyle never lost his strong Yorkshire drawl, and was as plainspoken as they come. He had little time for subtlety and rarely showed concern for others' feelings.

THE LITTLE GREEN MEN INCIDENT

A good insight into Hoyle's personality and approach can be gained by the furor that he sparked over the discovery of the astronomical phenomenon called the pulsar. In 1974, the Nobel Prize in Physics had been awarded to Professor Anthony Hewish, of the radio astronomy department at Cambridge. Hewish shared his prize with department head Martin Ryle, for general developments in radio astronomy, but Hewish's prize was specifically for the discovery of the pulsar.

Pulsars are very dense stars—neutron stars—that rotate at immense speeds, sending out pulses like a radio wave version of a lighthouse. These pulses can come every few seconds or as frequently as a couple of thousandths of a second apart. It was Thomas Gold, whom we will soon meet developing an alternative to the

Big Bang with Fred Hoyle, who realized just what a pulsar was, but when Hewish and his PhD student, Jocelyn Bell (now Jocelyn Bell Burnell), made the discovery of these strange bodies, there had been nothing to compare with them. Here was a regular, pulsing radio signal from the stars.

Bell and Hewish originally called the signal discovered in July 1968 LGM-1, where LGM was a tongue-in-cheek reference to the possibility that a regular signal might be a message coming from an intelligent life form, "little green men" as aliens were often called at the time. Although they were in the Swinging Sixties, academia was still very stuffy, and it was suggested at the time that Bell and Hewish never seriously considered an intelligent extraterrestrial source for the signal, but the news spread fast because of this twist. In truth, it seems very unlikely that they didn't even wonder if that signal might indeed be our first real evidence that there is alien life out there.

The controversy that Hoyle stirred up, however, has nothing to do with any thoughts of alien involvement. When Hewish received his Nobel Prize, Bell was not mentioned. There's an interesting parallel with the 1962 Nobel Prize in Medicine, which also caused controversy. Here the prize went to Crick, Watson, and Wilkins "for discoveries concerning the molecular structure of nucleic acids and its significance for information transfer in living material," discovering the structure of DNA. In this case another female researcher, Rosalind Franklin, was excluded. But here the Nobel committee had an excuse in the form of the rules. The prize couldn't be shared by more than three people, nor could it be awarded posthumously, and Franklin had died by the time it was awarded.

Neither of these reasons applied in the case of Bell, and in 1975, when the prize was announced, Hoyle steamed into the situation in

Bell's defense. In a press interview before a series of lectures (not about pulsars) at McGill University, Montreal, Hoyle was asked what he thought about the circumstances of the discovery of the pulsar. He suggested (in what he later described as "a conversational style expecting the reporter in question to follow what I have found to be the normal American practice, that controversial material is checked back before publication") that Hewish did not deserve the credit.

According to Hoyle, Bell made the discovery, but then was joined and "swamped" in the paper on the subject published in *Nature* by the directors of her group. (Bell herself denied being squeezed out.) Hoyle pointed out that if they had followed normal practice and published the discovery first, rather than keeping it secret and only publishing after a follow-up investigation led by Hewish, Bell would have gotten the glory for the initial discovery, but as it was, it was Hewish's follow-up that distracted the Nobel Prize committee. This forthright support (even though Bell didn't seem to really want it) was typical of Hoyle's approach.

THE STEADY STATESMEN

Along with two colleagues from Vienna, also working at Cambridge, Thomas Gold and Hermann Bondi, Hoyle proposed that the universe was in a steady state of continuous creation and expansion. Although this theory had good scientific reasoning behind it, as did a number of other alternatives to the Big Bang, it seems partly to have emerged from a dislike of the Big Bang because the suggestion of a beginning of time and space was too like the biblical creation. As Stephen Hawking has put it, "Many people do not like the idea that time has a beginning, probably because it smacks of divine intervention." This was certainly the case with Hoyle, who was as vocal an atheist then as Richard Dawkins is now.

According to the Steady State model, as the universe flowed outward, more matter came into being, keeping the universe as a whole in a constant condition. There was no beginning; there was no end; the universe simply went on forever, always growing, always staying the same as more matter came into being throughout all space.

This wasn't just a matter of kicking out at the Big Bang for no reason. Before Steady State, the choice of theories was between the Big Bang or a static universe. All the evidence from Hubble's observations and the research that followed suggested the expansion of the universe was real, so if Big Bang were to have a challenger, it would have to be a theory that incorporated this expansion effect. And it seemed to Hoyle that the Big Bang should be challenged, in part because it opened up the uncomfortable question that this book addresses—If there was a beginning to the universe, what came before?—and in part because the best measurements of the day gave the oldest stars an age of around 20 billion years, but didn't allow for an expanding universe that had been around for more than 10 billion years.

This continuously flowing nature of the Steady State model meant that it was likened by its proposers to the then very popular 1945 movie *Dead of Night,* a personal favorite with the three scientists. One of the best supernatural films ever made, it's a compendium, with a series of short stories linked by people who meet up at an old house in the country. The movie ends with the events immediately before the first scene, so the whole is circular, without beginning or end, something that was particularly obvious in the days when movies were shown continuously, rather than with well-separated start times.

According to Hoyle, *Dead of Night* was the actual inspiration behind the Steady State theory. He describes Gold asking on the night the three of them had been to see the film, "What if the universe is like that?" It opened their eyes, Hoyle says, to the idea that something can be dynamic, yet at the same time unchanging, like a smooth flowing river. Thomas Gold, who first suggested the idea, has since disagreed with Hoyle's memory of the birth of the theory, saying that they only ever said the idea was "like *Dead of Night.*" Either way, the movie proved a good mental picture for them of something that could develop yet still end up in the same place.

In fact, the circular *Dead of Night* isn't the ideal way to think of the Steady State theory. Steady State is more like a chocolate factory with a 24/7 production line. Chocolate bars flow down the conveyor belt and out of the factory, out into the real world. The "chocolate universe" is expanding from the factory into the world, just as we believe the universe is expanding. But that doesn't leave an empty conveyor belt, because new materials are constantly being pumped into the system, producing more chocolate bars to flow down the belt.

CREATION IN THE GAPS

In the Steady State theory, the universe is constantly expanding, as the Hubble data had suggested, but instead of it thinning out, with bigger and bigger gaps between galaxies, as in the Big Bang picture, new matter is constantly being created in the gaps; eventually, over the vast evolutionary timescale of the universe, this matter condenses and forms new galaxies.

The obvious questions that the theory provokes are Where does all this new matter come from? and How does it know where to appear? Steady State supporters were ready for the first argument. After all, if you believe the Big Bang theory, every bit of matter in the universe was supposed to have come into existence at a moment in time with no particular reason for this occurrence. It was hardly more surprising that matter should trickle into existence constantly.

Hoyle imagined a hypothetical "C-field" that was responsible for the creation of matter, but this was just a device; there was no particular evidence for this field. He was just saying, "Matter comes into existence, and the C-field is what makes it happen." This is a bit like saying, "The gravity that keeps us on the Earth is caused by the gravitational field." It's a reasonable statement, but doesn't tell you anything new about gravity. Unlike the Big Bang which had to occur at a specific time, and needed an explanation for that too, the Steady State didn't need an explanation for "why now?" because it was always happening. That's just how things were.

As for the second question, How does the new matter know where to appear? again, the Steady State theory seemed logical. Even Big Bang supporters accepted two key things about the universe. In fact these assumptions were necessary for Einstein's general relativity equations to work in the first place. The universe had to be homogeneous and isotropic.

HOMOGENEOUS AND ISOTROPIC

These terms provide a classic example of scientists using impenetrable words for perfectly simple concepts. An isotropic universe is one that is the same in whichever direction you look. The homogeneous part simply means that it is roughly the same throughout, so where you do that looking from really doesn't matter. Whatever measurement you take, it's pretty much the same anywhere, whether it's the speed of light or the density of matter. Of course, in practice there is huge variation in density between (say) near-empty space and a neutron star, but if you average out over these little local variations to the scale of galactic clusters, the universe seems to be pretty consistent.

Note that this essential pair of assumptions isn't a matter of certainty, even though there is no good reason for thinking differently. Now if the universe is homogeneous and isotropic, the answer to the second question about the Steady State model becomes fairly self-evident. Matter is going to come into being in the gaps because that's where there's room for it to come into being, and it will come into being all over, because the universe is homogeneous and isotropic. What's more, Steady State overcame a serious concern that still applies to the Big Bang model.

Although we can't apply general relativity to the universe unless we assume that it is homogeneous and isotropic, the Big Bang theory says it isn't isotropic in one particular direction: time. On the time dimension, in the Big Bang theory, the density of the universe has been rapidly decreasing as time passes. The past isn't the same as the present. Now if density is anisotropic with time (it varies as we move through time), why shouldn't other fundamental aspects of the universe vary too? Perhaps, for example, the speed of light was different when the universe was fresh (a theory later given much more detail by João Magueijo, who has suggested the speed of light was very different in the early time of the universe). If this is the case all our assumptions when we look back over time go out the window, and the fragile basis for our indirect measurements on the history of the universe is shattered.

Even so, to those familiar only with the Big Bang or a static universe, Steady State was initially quite a shock. When Fred Hoyle became its most vocal champion, he would defend it in a similar way to how Richard Feynman defended quantum theory (and specifically quantum electrodynamics, the unrivaled theory of how light and matter interact). Compare Hoyle's remarks about Steady State:

> [The Steady State theory] may seem a strange idea and I agree that it is, but in science it does not matter how strange an idea may seem so long as it works.

with Feynman on quantum electrodynamics thirty-three years later:

> It's a problem physicists have learned to deal with: They've learned to realize that whether they like a theory or they don't like a theory is *not* the essential question. Rather, it's whether or not the theory gives predictions that agree with experiment. It is not a question of whether a theory is philosophically delightful, or easy to understand, or perfectly reasonable from the point of view of common sense. The theory of quantum electrodynamics describes Nature as absurd from the point of view of common sense. And it agrees fully with experiment. So I hope you can accept Nature as She is—absurd.

When Gold threw the idea of Steady State into the discussion, both Hoyle and Bondi were initially dubious. They thought it wouldn't hold up to close scrutiny, but then quickly discovered that they could not knock a hole in it. The more they examined the idea, the better it seemed as an opponent to the Big Bang.

WHERE IS THE NEW MATTER?

Apart from the questions I posed above, the other big concern about Steady State was where that new matter was. Why can't we see it? This is a bit like people saying, "If evolution is the way new species form, why don't we see it happening all around us? Why aren't animals and plants evolving away before our very noses?" The trouble is that evolution happens over such a large timescale, that in any period we can observe there is very little to see. Similarly, if the Steady State theory were true, it would happen on such a scale that there would only need to be a small amount of matter created in our observing timescale. As Fred Hoyle put it, it would be the equivalent of an atom being created once a century in a space the size of the Empire State Building.

From the late 1940s when the Steady State theory was first envisioned, right through to the early 1960s, there was little reason to make an educated choice between the two theories, so the result was often a decision made on prejudice or association. Some represented the Steady State as more old-fashioned, because it was less different from the once-prevailing static universe theory. Others thought that Big Bang was old-fashioned because it harked back to Genesis in the Bible. Some divided on national lines, as Big Bang had come out of the United States whereas Steady State was a British theory.

Although less remarked on, there were interesting parallels between these theories and the development of geological theory a century before. As we have seen, it had been assumed for a long time that the Earth reached its form in a series of catastrophes, such as the original creation and the great flood described in the Bible. But as geology became more of a science and less of a cataloging exercise it was realized that it was much more often a gradual process, described as gradualism or uniformitarianism. In this style, broadly the approach accepted today, geological processes are assumed to continue as they always have, without sudden catastrophic change. Steady State was a uniformitarian theory, whereas Big Bang was a catastrophist theory, the kind that was largely dismissed from the philosophy of science in fields other than cosmology.

Once Steady State was well established, it had a lot going for it, not so much because it was possible to show that Steady State was right, but rather because a key element of Big Bang didn't fit with reality. Remember, it is only possible to disprove theories, not to prove them. Big Bang's big issue was the age of the universe. As we saw on page 68, all the way from the 1930s onward there was a real problem that the age of the universe that came as part and parcel of the Big Bang theory appeared to be significantly younger than the Earth and the stars. It was only when that problem was overcome that the Big Bang had any chance of success.

By the mid-1950s, though, the age of the universe was extended (see page 79) first to 3.6 billion years, then over 5 billion, and eventually past the 10 billion mark. There was no longer an obstacle in the way of Big Bang in terms of the age of the universe, but in itself this didn't sound the death knell for Steady State. Just because Big Bang was now logically feasible didn't make it true.

STEADY STATE BEGINS TO CREAK

For a good number of years the Steady State theory gave the Big Bang a run for its money, but over time, better observations suggested that the universe had undergone very significant changes in the distant past, something not allowed for by the original Steady

State theory. The farther out you look into space, the farther back you see in time, because the information travels toward us at the speed of light. This means we can see back billions of years, if we have good enough instruments, and since Gold's original formulation of Steady State, these instruments have gotten better and better.

Another factor is more theoretically based. The Big Bang theory assumed much of the helium (the second heaviest element after hydrogen) was created in the aftermath of the Big Bang, where the original Steady State theory assumed that this helium was created in stars. We know this happens. Stars such as our Sun depend on the conversion of hydrogen to helium to generate the nuclear energy that powers them and keeps us alive. But with the assumptions for the lifetime of the universe based on Big Bang–centered theories, stars couldn't produce as much helium as we see, and you'd expect it to be distributed more around the stars than across space as it actually seems to be. (Hoyle would later argue that the universe could be significantly older, giving time for the helium to be produced in the stars.)

RADIO MAKES WAVES

Next, a very different type of telescope would come to figure importantly in the great cosmological debate. By the mid-1960s, astronomy had taken on a much wider range of instruments than the optical telescopes that had remained unchanged, except in scale and subtlety of design, since the days of Galileo and Newton. Light comes in a whole range of wavelengths or energies, from the ultrahigh-energy gamma rays through X-rays, ultraviolet, visible light, infrared, microwaves, and radio. Back in the 1930s and 1940s, engineers working on radio receivers and radar had accidentally discovered that there were radio sources out in the universe. The Sun, for example, was a source of radio just as much as it was of visible light.

It was gradually realized that these nonvisible light outputs could provide an alternative to the traditional telescope that would make it possible to see different cosmic objects, or existing ones in a new way. Radio telescopes are much less precise than optical telescopes, because radio waves have much longer wavelengths (the photons have less energy), yet it is possible to pull together radio signals across a much broader receiver. Where the biggest optical telescope had a two-hundred-inch mirror, radio telescopes were soon being built with a two-hundred-*foot* receiver, and virtual telescopes miles across were possible by linking a number of receivers together.

Oddly, the first great success of radio telescopes would support a theory from the Steady State side of the Big Bang–Steady State argument, but this discovery would later unsettle the foundations of Steady State itself. When the radio astronomy group at Cambridge headed up by Martin Ryle first started mapping radio sources, they assumed those radio blips were coming from stars in our galaxy. But Gold and Hoyle were convinced that these sources where nothing had been visible before were distant galaxies.

Confirmation from the big optical telescopes showed that the new sources were indeed galaxies, but Gold and Hoyle's success was used against them as these radio-emitting galaxies proved largely to be young galaxies, and they were all far out in time and space, where Steady State expected young galaxies to be evenly distributed throughout the universe. However, this initial triumph over Steady State was short-lived. When Ryle produced his initial results at a lecture at Oxford on May 6, 1955 (the day I was born), he was using data that did not stand up to scrutiny. Another set of results published by Australian radio astronomer Bernard Mills just two years later showed that Ryle had been premature in declaring the death of Steady State.

Mills said, comparing the Australian catalog with the Cambridge one,

> We have shown that in the sample area which is included in the recent Cambridge catalog . . . there is a striking disagreement between the two catalogs. . . . We therefore conclude that discrepancies, in the main, reflect errors in the Cambridge catalog, and accordingly deductions of cosmological interest derived from its analysis are without foundation.

Mills was saying that it was impossible to draw any conclusions on the validity of Big Bang or Steady State from these observations of radio galaxies.

Things were made significantly worse for Hoyle, however, when Ryle came back with a much more detailed survey which pointed in the same direction as his first results, if more weakly. Rather than go through the usual scientific publication channels, Ryle got the Mullard company, who funded much of the radio astronomy work at Cambridge, to ask Hoyle to sit in on the platform at a press conference to discuss some interesting results. Hoyle was not told in advance what this was about, but as soon as Ryle started, he says that he "realized he had been set up."

Ryle announced that his new results showed that the Steady State theory was wrong. Would Hoyle care to comment? Ryle apparently realized he had gone too far, putting Hoyle on the spot like this, and later rang to apologize, saying, "He never realized how bad it would be when he had agreed to the Mullard suggestion of giving a press conference." This certainly isn't a conventional way to announce academic results, nor to allow a colleague to respond to them. But Steady State had taken a very public blow, and even if there hadn't been further supporting evidence it was probably already doomed by Ryle's clever use of public exposure.

KINDERGARTEN GALAXIES

As instruments, both optical and radio, improved, it became more and more apparent that there are features when you look back in time by looking into the far distance that don't seem to exist now: quasars are a perfect example. Quasars seem to be the answer to a question Hoyle himself posed in the early days of the Steady State theory as a way to show which of the theories was right. Both theories explained pretty well every astronomical observation that had been made by the 1940s, so Hoyle dreamed up a test that should distinguish between them.

The Big Bang theory is a bit like a single class of children passing through school. They begin in kindergarten, go through elementary school, junior high, high school, and finally college. At any one time, the whole class is in the same grade. The Steady State theory is more like a whole school system. At any one time, there are children in all the different grades, some just beginning, others mature. Swap children for galaxies. In the Big Bang all the galaxies should be roughly similar in age. They are all mature, grownup galaxies. To see those "kindergarten" galaxies, you would have to look far back in time (which for astronomical purposes, given the limited speed of light, is the same as saying look far out into space). In the Steady State, there should always be new galaxies forming, so there is no need to look out into the far depths of time and space for youngsters. They should be right here, on our doorstep.

In the late 1940s, astronomical equipment was just not good enough to see whether these "kindergarten" galaxies existed. If they were nearby, as Steady State suggested, the telescopes of the time would not be able to distinguish them from grownup galaxies, and if they were far distant, as Big Bang suggested, they would be well beyond the range of any telescope. Hoyle had proposed a test, but one that couldn't be carried out. Now, though, we can look much farther back in time and space. Quasars (short for "quasi-stellar objects") tend to look like a star with a jet shooting out of the side, but the appearance is deceptive. They are, in fact, thought to be huge young galaxies. And we only see quasars in the far, far distance, mostly more than 3 billion light-years away, well into the universe's past. There don't seem to be any "kindergarten" galaxies anywhere else. And that's bad news for Steady State.

QUASI-STEADY STATE

Hoyle did not go down without a fight. As late as 2000, just a year before his death, Hoyle published a book with colleagues Geoffrey Burbidge and Jayant Narliker that sought to go, as its subtitle said, "from a Static universe, through the Big Bang toward Reality." Hoyle's book is important because it identifies some genuine problems with the way cosmology had developed toward the end of the twentieth century.

As he pointed out, when experiments such as the COBE satellite (see page 146) were justified, requiring millions of dollars and the cooperation of agencies such as NASA, "Before the instruments can be built, extravagant claims must be made about what we shall find. Not surprisingly, when these instruments eventually work, we get a succession of even stronger claims to the effect that what was expected *has indeed been found.*"

"In an era when serious scientists can take such an approach," says Hoyle, "there is no room for the discovery of phenomena which are not already expected." The result, he suggests, is conformist science where it is very difficult to break out of the current accepted wisdom, whether it is about the Big Bang or other dominant but unproven theories such as string theory. Although Hoyle did have a tendency to unnecessary belligerence, as was seen when he waded in on behalf of an uncomfortable Jocelyn Bell, he does have a point.

When experiments are so large-scale, so expensive, there is a strong tendency to only fund work that supports what we already believe, rather than truly opening up new fields. It can be difficult to fight against the flow. And for that reason, if no other, it is worth giving Hoyle's last assessment of where Big Bang went wrong a brief look.

Hoyle and colleagues came up with the "quasi–Steady State model" which addressed pretty well all the problems that Steady State faced. In it, rather than matter being created in a single moment, an infinitesimal time after the Big Bang itself, matter is continually being created throughout space by "near-black holes." These are stars that are almost black holes (see page 203 for details on black holes), but haven't quite enough mass, so are still able to communicate with the outside universe. Such a model could be oscillatory (expanding and contracting over huge time periods) or expand indefinitely; Hoyle chose to use an oscillatory model, which provides a better match to other observations.

The quasars, which were used with such effect against the original Steady State theory, are in this model, not kindergarten galaxies, but spawned from existing galaxies. Hoyle points out that there are some concerns about the way red shift has been used to measure the extreme distance of the quasars, and Hoyle believed that they were much closer than first suggested, often linked (sometimes visibly so) with a galaxy that has emitted them as part of the creation process. As for the cosmic microwave background radiation (see page 61) which some Big Bang supporters take as virtual proof of its existence, Hoyle argues that this can be generated by galaxies and the creation taking place from near-black holes. He suggests that his theory predicts the observed energy of the radiation more precisely than Big Bang, which initially predicted a figure that was significantly different from the actual level.

Although it should be stressed that the continuous creation required in Hoyle and colleagues' model hasn't been observed, much of the mechanism of the Big Bang hasn't been observed either. What Hoyle has done here is a similar patching-up job to that done on the Big Bang to make it fit observed results. It is no more or less a problem for the quasi–Steady State model than it is with the current Big Bang model. (Compare this, for example, with the bouncing branes model on page 208 which has yet to require any significant patching up.) Hoyle makes a very valid point that the Steady State model has been sidelined and ignored because of the way in which big science works, not because of specific issues with the model.

The fact remains that outside the Hoyle camp, as problems arose from more detailed observation of the distant past in space, Steady State variants fell more and more out of favor and the Big Bang took on huge momentum. Steady State would never recover. Meanwhile, back in the late 1970s, the Big Bang theory was to face its own problems. But rather than developing an entirely new theory, its supporters looked for a way to patch up the theory to match reality. And where Hoyle's attempts to revive a failing theory were considered ineffective, here the idea of a Band-Aid for the concept was to be welcomed with little criticism. Big Bang with added inflation was about to be born.

7.

INFLATING THE TRUTH

> When proposing a new idea experience shows that it is a good maneuver to begin by identifying new weaknesses of the existing paradigm and then to show how the proposed new idea removes them. . . . We believe the success enjoyed by the inflationary model rests not so much on its intrinsic merit, but on how it was presented to the cosmological community.
>
> —FRED HOYLE, GEOFFREY BURBIDGE, and JAYANT V. NARLIKAR,
> *A Different Approach to Cosmology*

As the Steady State theory faded away, the Big Bang came into more prominence, but with the growing realization that there was something horribly wrong with it. If the universe had expanded the way it was thought to have, there just wasn't enough time for it to have become as uniform and flat as it actually appeared to be. It was a young physicist called Alan Guth who devised a bizarre solution to this problem. Born in New Brunswick, New Jersey, in 1947, Guth attended MIT and moved on from there to Cornell, but it was at Stanford that he would offer the solution to the biggest problems the Big Bang faced.

TURNING BACK GRAVITY

Guth suggested that soon after the Big Bang, for a very short period of time, gravity effectively reversed. Instead of pulling things together, it pushed things apart at immense speed. Soon after, it switched back to normal, and the universe continued to expand in the way it is doing now. Guth had no good explanation for why this should have happened, but *if* it happened, it would explain why things are so different from expectations.

All the scales involved here are dramatic. At the Big Bang we start with something totally outside our physical understanding, a singularity, a point of zero size with infinite temperature and energy. Because this isn't accessible to our math, it's possible it didn't quite start at a point, but we are certainly talking something crammed with mass and energy in a tiny space, something that appears out of nothing for reasons that are no better explained than they are by Genesis.

Just 10^{-35} seconds after the Big Bang, inflation began, for no known reason. (This is a slight oversimplification: there are now explanations for inflation involving the breaking of symmetry in the initial force that would eventually become the familiar forces such as electromagnetism, and involving phase transitions rather like the change from water into ice, but many would argue these ideas don't satisfactorily explain why inflation occurred when it did, or stopped when it did.)

Things had hardly got started at this point. 10^{-35} seconds is an inconceivably short time, like $\frac{1}{10}$ of a second, but with thirty-five zeroes at the end instead of just one. In that ridiculously small period of time, the universe was thought to expand by a factor of 10^{30} or more. It ended up more than 1,000,000,000,000,000,000,000,000,000,000 times bigger. That didn't make it anywhere near the size that the universe is today. Even after inflation it was only around a kilometer across. But before that it had been ridiculously small.

The result seems to contravene Einstein's special relativity that states nothing can travel faster than light. If anything could do so, it would travel backward in time with potentially disastrous consequences for causality and the whole of reality. But as with the more routine expansion of the universe, Einstein's limit does not apply as it refers to the speed of something traveling through space. Here nothing was traveling. It was space itself that was expanding, so inflation caused no movement compared with the speed of light within the universe.

Because the initial size of the universe involved at the time of inflation was so small we'd expect quantum effects to come into play as the inflation (somehow) turned off and the inflationary energy was dispersed throughout what had become a very flat, uniform structure during that incredible stretching process. However, as we are dealing with quantum processes, uncertainty comes into play, and this means that there would not be a totally uniform result (just as well, to be honest, as without that variation there would be no seeds from which the stars and galaxies could form).

NOTHING IS CERTAIN EXCEPT UNCERTAINTY

Uncertainty is an important word in the quantum world. Heisenberg's uncertainty principle, one of the key components of quantum theory, says that there are pairs of properties which are linked in such a way that it is impossible to know both in absolute detail. The more you know about one, the less you know about the other. One of these pairs is momentum (mass times velocity) and location. If you know the momentum of a quantum particle exactly (what it weighs, its direction, and how fast it is traveling), you cannot know anything about its position. It could be anywhere in the universe. Similarly, if you know exactly where a particle is, it could have any momentum.

This is why quantum particles always jiggle around: the higher the temperature, the more they dance about. If a particle were absolutely still we would know exactly where it was and it would become a bizarre particle that could have any momentum. This is a really important aspect of quantum theory, so it is worth exploring it a little more. When Werner Heisenberg first came up with the uncertainty principle he misunderstood one key aspect of it, and that misunderstanding is still often presented today.

When Heisenberg first told his boss, Neils Bohr, about the uncertainty principle, he put it across in the form of an imaginary microscope. He described a particle as an electron passing through a make-believe ultrapowerful microscope. We use light to examine the object, so a beam of photons (quantum particles just as the electron is) is constantly crashing into the electron. The result is that the electron's path is changed. You can't look at a quantum particle without changing things.

Heisenberg is said to have been reduced to tears when Bohr ripped his idea to pieces. Heisenberg had assumed that until the microscope scanned the electron, the electron had an exact position and momentum. He thought it was the process of observing it that messed things up. But actually, Bohr pointed out, the uncertainty was more fundamental than that. There was no need to observe the electron for uncertainty to apply: it was inherent to the nature of a quantum particle.

Uncertainty also applies to the fields such as electromagnetism that are part of quantum theory. This means that particles can pop into existence out of nothing, briefly exist, and then disappear again. This kind of activity is what is envisaged as producing the quantum fluctuations after inflation, with tiny fluctuations in the inflation itself producing regions with subtle variations in density of energy.

THE PHANTOM HISS

The quantum variation in the stretched remnants of the Big Bang is where the best evidence so far comes from that inflation (and the Big Bang itself) exists. It all started with some unwanted pigeon droppings back in 1965, when two researchers were using a telescope that looked like a strange ship's funnel lying on its side, to scan the skies for interesting radio signals.

This special radio receiver was originally designed to pick up signals from a short-lived reflective satellite, and was still being employed in 1965 to communicate with the primitive Telstar satellite, but was also by then being reused as a telescope. The researchers were Robert Wilson and Arno Penzias, based at Bell Labs in Holmdel, New Jersey. Wilson, born in Texas in 1936, and Penzias, born in Munich, Germany, in 1933, and at age six, evacuated with other Jewish children, were rare radio astronomers among the communications engineers at Bell Labs. They were looking for signs of a cloud of gas surrounding the Milky Way galaxy. But what Wilson and Penzias found was a background hiss, a signal that seemed to come from everywhere.

To begin with, their response was to assume that the source of this noise was earthbound. It's not uncommon for those who use radio telescopes to be misled by a powerful local source, whether it's a radio ham, power lines, or a badly wired vacuum cleaner. It's much easier to get radio interference from the ground than it is for a conventional telescope to get light interference. But Wilson and Penzias eliminated these causes. Wherever they pointed the telescope, even at New York, there was no variation in this background hiss.

They also checked for faults in the telescope itself. There was nothing wrong with the wiring or electronics that could be producing the signal, but they discovered that the local pigeons had been using the wide horn of the telescope as a perch. Its opening was coated with white bird droppings (or "dielectric material" as Penzias coyly described it), because a pair of pigeons had decided to nest there, despite the regular disruption to their lives as the telescope rotated and flung them around.

Penzias and Wilson, with perhaps unusual scruples, bought a humane pigeon trap and sent the birds forty miles away, the location chosen because it was the farthest they could send them using the internal mail (presumably not in a reusable mailer). Unfortunately their efforts were for nothing. In a couple of days the aerial pests were back and had to be shot. When the droppings were cleaned away, the result was not to eliminate the hiss. It was still there, an everpresent noise beneath the more sharply defined radio sources that were galaxies and pulsars.

Penzias and Wilson did not know what they had found, but Penzias happened to phone another radio astronomer to discuss something completely different. He only mentioned their problems with the unwanted noise as a passing remark. But Bernie Burke, the man he had called at the Carnegie Institution in New York, had heard about the work that another scientist, Robert Dicke, and his group were undertaking over at Princeton University.

REMNANTS OF THE BIG BANG

Missouri-born Dicke was better established than the Bell pair, already in his late forties, and heading up a team at Stanford. He had his own theories about the beginnings of the universe (see page 188) which like the traditional Big Bang theory involved a very hot initial soup of matter and radiation. In a sudden moment of inspiration, Dicke believed he had thought of a way of looking back to the very beginning.

Unlike Gamow's team, who had predicted there would be radiation from the Big Bang, but thought it would not be detectable, Dicke realized that the remnants of that initial wild radiation should still be visible today. Dicke and his small team were actively looking for the radiation that started flowing when the universe first became transparent. Dicke had worked very successfully in his early career on general relativity, but during the war had been diverted (like so many of the cosmology mafia) into radar research, which probably inspired this realization.

Dicke's two young research assistants collected old war surplus equipment; luckily the microwaves they expected to remain from the Big Bang were very similar to the radar signals that had been used in the Second World War, making it possible to reuse many of the parts that were available very cheaply now they were surplus. They put together their makeshift telescope up on the roof of Princeton's Guyot Hall, high enough, they hoped, to keep it away from the manmade influences around them.

It was bad enough that the Princeton team was working with such make-do equipment, but they also faced a problem that the other practitioners in their new but rapidly widening field of radio astronomy didn't face. Try though they might, there was no way to get every bit of background radiation out of the system. Some came from the atmosphere. More still came from the equipment itself. Regular radio astronomers got around this by comparing an "empty" bit of sky with something they were observing. The difference between the two showed what the real signal was. But the cosmic background radiation was supposed to be there wherever you looked.

With no way to point their Rube Goldberg telescope at some background-free sky, the Princeton team resorted to building an artificial source that would produce the kind of signal you would expect from a known energy of background radiation. They could then use that to calibrate the results from the telescope, and if they found something that stayed the same across the sky with an energy in the appropriate range, they had a fair chance of identifying it as the cosmic background. There was only one other radio telescope in the world that had this ability to make such absolute measurements rather than a comparison. The one Penzias and Wilson were using.

It was only when Burke, the man Penzias had called, got in touch with Dicke that he realized that what the New Jersey telescope seemed to have picked up by accident was what his group had been painstakingly searching for: the cosmic background radiation, that mysterious "echo" of the Big Bang. As we saw earlier, this is not really an echo, but is thought to be the actual light from the point when the universe became transparent, brought down to the energy of microwaves by the red-shift effect of the expanding universe. Dicke and his team drove over to meet Penzias and Wilson, and among them they pulled together data and theory to be able to consolidate their ideas of this early time in the history of the universe.

THE FORGOTTEN FORERUNNER

When George Gamow first heard of the two papers written by the teams (one on the discovery, the other on the implications) he was probably delighted. Here was vindication of his idea. People would realize what a great concept he had formulated. However, when he read the papers he was horrified to discover there was no mention of his work, no references to his own papers. Just as Gamow had accidentally pushed Alpher and Hermann out of the limelight, so he himself had now been forgotten.

Even today, with papers much easier to access in electronic form, there is still a possibility of being unaware of earlier work. Although researchers will do a search to find prior references, and should be knowledgeable of the literature, it is quite easy for old research that was thought to be a dead end at the time to be forgotten. Back then, it wasn't surprising that Gamow's ideas, which had been cast aside when he was shown to be wrong in thinking that heavy atoms were made in the Big Bang, weren't particularly familiar to Penzias and Wilson or to Dicke's group.

Part of the problem is the tendency of scientists to specialize. The radio astronomers were only concerned with microwave work, and only read journals for those working in this particular field. Gamow had written both for a popular audience and for physics journals. There is no reason to suppose Dicke was lying when he said he had never come across Gamow's ideas, famous though the physicist was.

Another example of this lack of ability to pull things together across disciplines is the surprising fact that the background radiation had been already observed indirectly when Gamow's team first predicted it should be there; it's just that no one realized it had happened and combined the observation with Gamow's theory.

This was an observation that predated radio telescopes, made by the astronomer W. S. Adams in 1938 at the Mount Wilson Observatory where Hubble had made his discoveries. Adams was looking at a star and found that its spectrum seemed to be subtly altered, as if it were being interfered with by passing through a gas made up of cyanogens, compounds of carbon and nitrogen (probably best known as a constituent of the lethal substances used in the gas chamber). Adams' observations only made sense if the cyanogen molecules were dancing about with the energy equivalent to a temperature of about 3 degrees K.

However, Adams knew that in empty space, the energy should be closer to zero. Although it is entirely possible for gas clouds to have significantly more energy, which they absorb from starlight, this wasn't what was expected here. What Adams seems to have detected is the effect of the background radiation on those gas molecules, stimulating them into motion. This would never have been a strong enough confirmation to support Gamow's theory alone: there were too many other causes for the energy, but had the connection been made, it would have been a good initial pointer that would have inspired searches for the background radiation much earlier.

The initial data from Penzias and Wilson's telescope were enough to suggest Gamow could be right, and this was the last nail in the coffin of the Steady State theory, which had no explanation at the time of why there should be this microwave background radiation. (As we have seen, Fred Hoyle would later argue there was a sensible explanation for the background radiation with the enhanced Steady State theory, but this came too late to save the concept.)

SEARCHING FOR THE RIPPLES

The discovery was enough to eventually win Penzias and Wilson, although surprisingly not Dicke, the Nobel Prize. But the Bell Labs telescope wasn't sensitive enough to give any concrete confirmation of the first discovery, nor to see if what was observed matched the expected variation from quantum effects during inflation. These variations meant that the cosmic background radiation shouldn't be entirely uniform, but should have highs and lows. Where there was more matter it should be a little lower in energy as the matter scattered the light more. After all, where did all those lumpy bits such as galaxies we now see come from if the original preinflationary universe were perfectly smooth and homogeneous?

By now Dicke's team were up and running with their makeshift telescope and were able to add one result to the information obtained thus far: that cosmic background radiation was pretty consistent wherever you looked. Although this sounds as though it's contradicting that hoped-for variation, at the level they were able to measure, those tiny variations wouldn't be visible. But before there was any chance of finding the fluctuations, it was essential to ensure that the background radiation was almost the same wherever you looked, and this Dicke's group was able to confirm, scanning the skies using the natural turn of the Earth to pass their telescope across different slices of the universe. All seemed uniform. This really was background radiation.

Getting to the quantum variation would mean building a receiver so sensitive that the atmosphere's effect, and particularly interference from water vapor in the atmosphere, could not be tolerated. Although balloons and high-flying aircraft were used to attempt to get around this, in the end it took a satellite to truly reveal the glory of the cosmic background radiation and its telltale ripples.

THE SATELLITE VIEW

This was the reason for those strange elliptical pictures of the universe that would later be produced by the COBE and WMAP satellites, and is why astrophysicists get so excited about a picture that really doesn't convey much to the general observer.

In those squiggles of light and shade we see a background radiation that fits well with both the concept of the Big Bang under Gamow's assumptions and a universe that had undergone incredibly rapid inflation in its infancy, with tiny variations that could provide the seeds for the massive local variations we now find.

The first of the two satellites, COBE (COsmic Background Explorer) was proposed in the mid-1970s, and finally launched in 1989. It should have been aloft sooner, but was originally scheduled for a launch on a conventional rocket; then NASA changed its policy requiring everything to be launched by the Shuttle, which forced a redesign. Sadly, the launch schedule was totally shattered by the *Challenger* disaster, so COBE was delayed for a number of years before a suitable launch vehicle could be found, requiring more redesign. Ironically, after considering the European Ariane launcher, the satellite's team went back to the original planned Delta rocket.

Two years later, COBE scientists were to publish the famous elliptical maps of the background radiation from space with psychedelic ripples across them. In fact the original picture was something close to a fraud in the way it was presented by the media. Those maps were more impressive looking than the results behind them. Most of the variation in the image, producing those impressive blobs and fringes, was from random radiation produced by the microwave detector itself. When this has been ironed out by comparing results at different wavelengths and using statistical analysis, there is very little left.

However, there were still the tiny expected variations, which from later, more detailed surveys would prove to be just a 1 in 100,000 deviation, a fluctuation that it was hoped was the signature of the remnants of the Big Bang. It was there, but it was just significantly less photogenic than the image that the newspapers and TV news carried.

There's another slight variation in the cosmic background radiation that doesn't appear in those pictures as it had already been factored out in the COBE maps. The Earth isn't stationary. We are circling the Sun, which rotates with our whirling galaxy, which is flying through the universe. Taken all together, we are moving at around 600,000 meters per second (around 1,342,000 miles per hour) against the backdrop of the radiation, which means we introduce a tiny red/blue shift into the radiation from our own movement, making the bit ahead of us slightly more energetic than the bit behind.

UNIVERSAL HYPE

Arguably, the media reaction to COBE was over the top, fueled as it was by comments from scientists who really should have known better. Perhaps the best-known living scientist in the world for the last twenty years or more, not hindered by his appearance on *Star Trek: The Next Generation,* has been British astrophysicist Stephen Hawking. It was Hawking who really stirred things up by referring to the COBE results as "the greatest discovery of the century, if not of all time."

There's no doubt that the COBE result was an important one for cosmology, but bearing in mind it was by no means detailed enough to be definitive, and was anyway such an indirect measurement that it could only ever suggest a particular theory was right, Hawking's comments seem overly enthusiastic. That same century had seen the discovery of the structure of the atom, quantum theory, relativity, and DNA to name but a few, making his claims more than a little exaggerated. But at least they had the rare effect of bringing an undoubtedly important scientific discovery to the eyes of many more of the general public than otherwise would have been the case.

It's also true that COBE's results were highly convincing to the scientists studying them, and showed (from a different detector onboard) a wonderfully clean blackbody radiation spectrum, suggesting that the early universe really could be considered a simple blackbody source with very limited structure. The results would be backed up in much more detail by the later WMAP satellite (the Wilkinson Microwave Anisotropy Probe), which by the early years of the twenty-first century was producing more detail on the

WMAP satellite image

actual distribution of energy among this early light, and also providing other parameters that enabled both the age of the universe to be much better estimated, and to give a better view of the makeup of the universe.

Looking at the output from WMAP as it is usually presented is a bit like taking a look at an elongated speckled bird's egg. It's a squashed oval with dark and light flecks all over it, sometimes shown in garish artificial colors. It's hard to see how anyone can deduce anything much from these uninspiring-looking blobs. But what you have to realize is that what you are seeing here is a huge condensation of what WMAP actually records. The WMAP satellite takes six months to scan the whole sky, all the while rotating to capture thin slivers. That squashed egg is an amalgam of those many slivers, but with nothing like the detail that is available to help scientists investigate this residual glow of the early universe.

Measurements that are possible now that we have these satellite-based microwave telescopes suggest the actual background radiation temperature is around 2.7 K (−450°F), making Gamow's prediction of 5 K a surprisingly good match to reality, although a cynic might say, "Well, you would expect space to be cold, wouldn't you?"

AFTERGLOWS AND ECHOES

It's still a little hard to grasp how the cosmologists understand that this extremely low temperature measurement is somehow an "afterglow" of the Big Bang. As we've seen, a typical explanation goes something like this. In the immediate aftermath of the Big Bang, the (relatively tiny) extent of space was opaque. The matter it contained was so hot that it was a superheated plasma containing lots of free electrons which would scatter and redirect all the photons, meaning no light could get through it. When it cooled enough, it became transparent. The electromagnetic radiation that was around then has been "circulating" ever since and it is this afterglow, this residual radiation, we see in the cosmic microwave background.

That's all very well, but as an explanation it is highly unsatisfying. If this is "afterglow," what caused the time delay from the original action to the current glow? This is typical of a description that oversimplifies reality to the point it is actively confusing if you think about it, rather than just take the vague feeling it gives. Aftershocks and afterglows are all very well as metaphors, but what is the metaphor describing?

The real picture is not of an aftershock circulating around the universe, echoing back and forth, or of an afterglow filling the sky after a big fire. Instead we are talking about light that was traveling through that newly transparent young universe that is still on its way because of the enormous expansion the universe has undergone. As the radiation was initially everywhere, wherever you are in the expanded universe, it should still be flowing past.

There's no doubt that detecting such radiation was a useful support for the Big Bang theory. But it is possible to go too far. In his otherwise excellent book, *Big Bang*, Simon Singh comments, "Anyone who could detect this so-called *cosmic microwave background radiation* (CMB radiation) would prove that the Big Bang really happened." Singh should know better. They would do nothing of the sort.

As we've already seen, you can't prove a scientific theory; you can either offer evidence to support it, or you can disprove it by providing evidence that contradicts it. The existence of cosmic microwave background radiation was useful supporting evidence for the Big Bang theory, but it is highly indirect evidence that could have many other reasons for existing and would later be used to support several opposing theories. There is no way that it can be regarded as absolute proof.

BIG ANNIHILATION

By the time the universe became transparent, there was another change that had to have occurred, a problem that worried cosmologists for a long time. Specifically they wondered why the early universe didn't explode in a vast matter/anti-matter collision.

When the universe first came into being, there was no particular reason why there should have been a preference for matter or anti-matter. What's more, in the incredibly high energy state immediately after the Big Bang, energy would constantly be converted into pairs of matter and anti-matter particles. In principle, there should have been equal amounts of matter and anti-matter, which then should eventually have wiped each other out, leaving a universe full of energy alone.

That this didn't happen is usually explained by assuming that very subtle differences in the properties of matter and anti-matter meant that there was a tiny extra percentage of matter; everything else then wiped itself out, leaving only this excess. This theory, devised by Andrei Sakharov, the Russian physicist better known for being a political dissident, suggests that as little as one particle in a billion survived the vast matter/anti-matter wipeout. But that was enough.

Some have speculated, however, that instead the universe in some way became segmented, and that there are vast pockets of anti-matter out there, perhaps on the same scale as our own observable universe. If the two ever came into contact, the result would be an outpouring of energy that would make every super-nova ever seen combined look like a match being struck.

Although the Big Bang is the best accepted theory, it certainly isn't the only one, and a number of scientists regularly pick holes in the evidence supporting it. It doesn't help that the whole thing has the feeling of something held together with a Band-Aid. If it hadn't been for the addition of the idea of inflation, the whole concept wouldn't work, and the trouble with inflation is that although it's something that would make what's observable possible, no one can come up with a decent explanation as to why it should have happened.

ADDING UP THE POSSIBILITIES

There is one possible argument for inflation, still under development at the time of writing, advanced by Stephen Hawking of the University of Cambridge and Thomas Hertog of Denis Diderot University (Paris, France). It uses a similar approach to the "sum over paths" solution produced by the great U.S. physicist Richard Feynman to explain some aspects of quantum theory.

Quantum theory depends on probability rather than certainty. The sum over paths approach says that a particle doesn't really take the obvious straight-line course from A to B. Instead its progress includes every single one of the infinite range of possibilities of paths. Many of these have low probability, or have phases (a state of the particles that varies over time) that cancel each other out, resulting in the expected path, but these paths are real, not just a mathematical trick to get to the solution.

Imagine, for a moment, a beam of light reflecting at an angle off the middle of a mirror. If you chop off most of the mirror including the middle, leaving only a piece to one side, you obviously won't get a reflection. But put a series of thin dark strips on the remaining piece, which only leave available those paths where phases add up, and it begins to reflect, even though the light is now heading off in a totally inappropriate direction for reflection as we understand it. Those strips stop the strange paths canceling each other out.

You can actually see this happening without bothering to fiddle around with mirrors and dark strips. The phases of photons change at different speeds depending on the energy of the light. In visible-light terms, that variation in energy is shown by changes in the color of the light. So different colors will be reflected off at an angle to a different degree by these fine lines. Shine white light on a special mirror with fine lines engraved on it and you should see rainbows. Practically everyone has a mirror like this: a CD or DVD. Turn it over to see the playing side and tilt it against the light. The rainbow patterns you see are due to the rows of pits in the surface cutting out some of the light's phases, leaving light reflecting at a crazy angle into your eye: a visible example of quantum mechanics at work.

Hawking and Hertog have suggested that the early universe at the Big Bang was a quantum object, and so can be treated using quantum mechanics. Just as the quantum particle takes every route between A and B, so they suggested, this quantum universe, which theory predicts has many, many possible forms (we're talking 10^{500}, that's 1 with 500 zeroes after it), was in all those different forms simultaneously, and the outcome was the sum across all quantum universes. They then selected for the sort of universe we have ended up with, looking at how a universe with matter and light behaving the way it does could fit into the different potential universes, and believing the result is that the sum across all the possible universes would drive inflation to be at the level that it needed to be to match current conditions.

There is a suspicion of a circular argument here. They selected for universes that could fit with what we've got, then applying them to the data results in . . . what we've got. But it should be emphasized that they started with the nature of the universe now and predicted inflation, which is only a theoretical precursor. If Hawking and Hertog are right, inflation is a natural result of the combination of possible quantum universes that they believe were necessary to form our own. It's by no means certain yet, but the likelihood of inflation is given a small boost by this.

Hawking and Hertog's work gives a theoretical reason why inflation might have taken place. However, it does have to be put up against another piece of theoretical work that may make it unlikely for the current ideas of inflation to continue to be held. The scientists doing this are using the very evidence that is usually held up as showing the Big Bang and inflation to have happened: satellite maps of the cosmic background radiation.

INFLATION, BUT NOT AS WE KNOW IT

This research emphasizes once again just how indirect the links are between what's observed and the theories that are based on those observations. According to a paper published in May 2008 by Benjamin Wandelt from the University of Illinois, Urbana-Champaign, the data do not support our picture of inflation. Although Wandelt is in a minority with his particular theory, many others feel that inflation is limited or even wrong. Michael Turner of the University of Chicago has commented, "It might last for ten years, but it won't last for ever."

Wandelt's reasons for doubting inflation are about uniformity. According to the current standard theory, inflation is responsible for the fact that space is largely uniform in energy, even though the farthest reaches of the universe are much too far apart for information ever to get from one side to the other to make that evenness possible. If inflation occurred, these areas would have been initially much closer together than without it, and so could share information before the inflationary phase.

Part of this uniformity is the expectation that the energy (hence temperature) variations we do see across the microwave background should have a normal distribution. This is the very common statistical distribution that when plotted out produces a bell-shaped curve, with the majority of occurrences scattered around the middle, and roughly evenly distributed tails to either side, growing smaller and smaller.

In practice, there seem to be more cold spots than hot spots on the background radiation map: the distribution seems to be skewed toward the cold end, suggesting there's something wrong with the current inflationary theory. As yet these data haven't been brought up to the sort of level of certainty scientists like. There's around a 99 percent probability of the numbers being correct, where scientists like to see something closer to 99.9999 percent to declare the data safe. Even so it's enough to get supporters of inflation worried.

This isn't the only problem that inflation faces. Bearing in mind what an immense expansion it would have involved, over such a short time, you would expect the fabric of space and time to be so shaken up that the echoes of it would still be flying around the universe today. When spacetime undergoes rapid change, the expectation is the generation of gravitational waves, fluctuations in gravitational effects that should travel around the universe as ripples do around a pond when a large stone is hurled into it. These gravitational variations in their turn would influence the wavelengths of the radiation that is flying around the universe, but despite much looking, this effect has yet to be spotted. It doesn't mean it's not there, but it is surprising that nothing has yet been found.

MISMATCHED LITHIUM

There are also other problems in matching the way the universe has behaved to the theory. One comes up in the amount of the element lithium made in the initial preuniversal soup. As do many elements, lithium comes in two "flavors" or isotopes. If you remember your periodic table from school, chemical elements are identified by two numbers, the atomic number and the atomic weight. The atomic number tells you the number of protons in the nucleus, or electrons in orbit around the atom (both the same). The atomic weight tells you the number of heavy nuclear particles (protons and neutrons) present.

When it first became clear what atomic weight was, it was thought there was something horribly wrong with the idea, as it is a measure of the number of particles, and whereas for some atoms this is fairly straightforward, for others the value doesn't seem to make sense. Nitrogen, for example, has atomic number 7, and atomic weight very slightly over 14, so it's easy enough to say it has 7 protons and 7 neutrons in the nucleus. Chlorine, however, has atomic number of 17 and atomic weight of 35.45. It seems to have around 18½ neutrons, which just can't be right.

The solution to this problem is the realization that atoms can come with different numbers of neutrons in the nucleus. In the case of chlorine there are stable versions with both 18 and 20 neutrons. As there are more with 18 than 20, they average out around the 18½ mark. These different versions are called isotopes. Lithium, too, comes in two stable isotopes, Lithium-6 with three neutrons, and Lithium-7 with four.

According to conventional Big Bang theory, a lot of the lithium in the universe was made before the stars were formed, although as with helium, a proportion is produced in stars. However, a big problem has recently emerged. The theory doesn't match observation. There seems to have been only one-third of the amount of Lithium-7 in the early universe that theory predicts, whereas there was 1,000 times as much Lithium-6 as there should have been.

As with all measurements reaching back to the very early days, we are reliant on some indirect observation and calculation. The actual values are produced by looking at very early stars. By peering as far as we can into the distant universe and using spectroscopic analysis (see page 44) it is possible to get an idea of the proportions of the different elements in very early stars. Without a chance to have been cooked up in the stars themselves, this balance should roughly reflect what things were like as the stars were formed.

To find what theory predicts involves a more subtle measurement, relying as so often is the case on the cosmic microwave background radiation. If the Big Bang theory is correct about how helium, lithium, and beryllium came into being, the amounts produced would depend on the ratio of particles such as neutrons and protons in the original mix (collectively known as baryons) to the number of photons. This ratio can be deduced from the small variations in temperature in the picture of the cosmic microwave background radiation produced by a probe such as WMAP. When that ratio was worked out and plugged in, it produced good results for helium, but ones that were way off, as already described, for lithium.

One possible reason for the discrepancy is that stars don't necessarily do what we expect them to do. It's possible, for instance, that those old stars being used to provide data on lithium levels actually reduced lithium levels, which would be good for the Lithium-7 figures, but not so helpful for Lithium-6. Others argue that the detail possible in spectral analysis currently doesn't allow for the sort of accuracy that has been claimed for Lithium-6 levels in the old stars, that it's just too difficult to tell the isotopes apart with such a faint source.

However, if the values are wrong—and that's the view of many astrophysicists—then there has to be something wrong with the predictions of what was produced by the Big Bang. Most effort is probably going at the moment into a suggestion that the particle reactions under way in the Big Bang soup were more complex than has been assumed, that there might have been particles involved we don't normally see in the current state of the universe. If that's true, such particles could be discovered as a result of the new Large Hadron Collider at CERN (see page 243). What some are brave enough to consider, however, is another interpretation, that this is yet another suggestion that the Big Bang theory is simply incorrect.

IS THIS ANOTHER LUMINIFEROUS ETHER?

Taken all together, the problems with the Big Bang plus inflation theory are not tiny negligible gaps between observation and theory; they are huge chasms you could drive a cosmological coach and horses through. Most cosmologists speak as if these problems have been solved, but the solutions are really huge fudge factors, reminiscent of the introduction of the "luminiferous ether" centuries before. Not only has inflation been added, it has also been necessary to squeeze in other unexpected components such as dark matter and dark energy to make the current model of the universe fit the observed data.

Some would argue that the fixes to the model introduced to make it fit reality are so extreme that they imply we need to start again from scratch. Let's just take a moment to remember the ether, as a warning of the dangers of ascribing too much reality to those cosmological fixes. The idea was simple. Although Newton was convinced that light was composed of numerous minute particles that he called corpuscles (little bodies), his rivals preferred the idea that light was a wave, moving along like ripples on a pond.

It was already known that the medium of one of our other key senses, hearing, depended on waves. Sound was a series of waves through the air. Similarly, it was thought, light was also a wave, and later experiments, notably those of polymath Thomas Young, would demonstrate that light did indeed have wavelike properties. Yet if light is a wave there is a significant problem: What is light a wave in?

If you think of a wave on the surface of the sea, it is made up of up and down movements of the water molecules. The water itself doesn't travel forward as a result of the wave, but the wave transfers energy forward. If you doubt that the water doesn't travel in a wave, bear in mind that the movement of the water at the edge of the ocean is not purely a matter of traveling waves. If you think of a simpler wave, sent down a piece of string tied to a door handle, it is clearer that the particles carrying the wave (in this case the molecules in the string) only move up and down, but the energy of the wave travels forward.

Light crosses the empty vacuum of space from the stars, so if it manages this, what it was a wave in had to be explained by those earlier scientists. This oddity could easily be demonstrated by sucking the air out of a glass bell jar with a ringing bell in the center. As the air was removed, the sound of the bell would fade away to nothing, with no air molecules to carry the sound energy to the ear. Yet the bell would not gradually disappear. It would still be seen as clearly however strong the vacuum in the jar.

Faced with this horrible hole in their theory, scientists applied the patch of the ether. The idea was that there existed a medium that filled all of space, in which the waves of light rippled. This "luminiferous ether" had to be a pretty remarkable substance. It was totally undetectable, except by its effect on light. You could neither touch it nor feel it. Unlike the air, there was no resistance to moving through it. It didn't suck out of a bell jar with a pump. Even stranger, the ether had to be infinitely rigid. Normally when you send a wave through something it gradually loses energy as the floppiness in the substance conveys the energy away from the wave's path. Before long, the wave has gone. But the ether allowed light to travel seemingly forever, certainly much farther than a wave could be carried by any known substance.

Surprisingly, scientists stuck with the prop of the ether for quite a while after it was proved unnecessary. As science showed that light was an interaction of electrical and magnetic waves, each creating the other, pulling each other up by their own bootstraps, there was no need for a medium. Even more dramatically, Einstein would suggest that light could be seen as tiny particles, photons, just as Newton had speculated, again doing away with the need for an ether.

The nail in the ether's coffin was an attempt by the first U.S. Nobel Prize winner in the sciences, Albert Michelson, with colleague Edward Morley, to determine how the ether was influenced by the Earth's passage through it. Try though they might with more and more sensitive experiments there was no effect. It was as if the ether didn't exist, something most physicists now take for granted.

So bear in mind the history of the ether as we come to another of the patches that proved necessary to make the Big Bang plus inflation theory work, patches that brought us the exotic, dangerous-sounding dark matter and dark energy. Some argue the parallel with the ether isn't accurate. In their book *Origins*, Neil deGrasse Tyson and Donald Goldsmith argue that the ether was a totally different affair. "While ether amounted to a placeholder for our incomplete understanding, the existence of dark matter derives not from mere presumption but from the observed effects of its gravity on visible matter."

With due respect to Tyson and Goldsmith, they are wrong. Theirs is an argument from the status quo. I am sure that supporters of the ether theory (which means pretty well every physicist in the nineteenth century) would have said the same about the earlier failed phlogiston theory that assumed burning involved giving off something (the invisible untraceable phlogiston) rather than absorbing something (oxygen). Dark matter is as much a placeholder as the ether. Victorian scientists would have said that the ether was something derived from the observed properties of light. That doesn't mean dark matter doesn't exist, but it is equally a concept about which we should exhibit significant concern.

DISCOVERING DARK MATTER

This dark matter was the earlier of the two dark phenomena to enter the cosmologists' vocabulary in an attempt to keep their model of the universe alive. A number of astronomers in the twentieth century observed behavior in galaxies that seemed to defy New-ton's laws, even with the subtleties added by general relativity. The first worrying data were discovered in 1933 by California Institute of Technology astronomer Carl Zwicky. Observing a cluster of galaxies in the Coma Berenices constellation, Zwicky detected relative motion that seemed crazy.

In the cluster, at least 1,000 galaxies are close enough together to have a significant gravitational influence on one another. Zwicky discovered that the outer galaxies of the cluster are traveling much faster than would be expected. So fast, in fact, that unless they are much heavier than they appear to be, they should be flying away from each other like peas placed on the edge of a fast-rotating disk.

This same effect, but on stars within a galaxy, was discovered by Vera Rubin of Georgetown University in the 1960s. Our own Milky Way galaxy has stars toward its edge that are dashing around too quickly for the gravitational force of all the known matter in the galaxy to hold them in place. There had to be some-thing more, something with extra mass to provide that gravita-tional attraction, but that was invisible.

At first glance, this isn't a big problem. Deciding how much a galaxy weighs (even our own Milky Way) is little more than an advanced guess. We can't specify the weight of every star, nor even an accurate count of the number of stars in the galaxy; we can only take a rough measure by counting stars as best we can in a patch of the Milky Way and multiplying it up to the whole. As-suming that educated guess isn't as far out as the motion of galax-ies suggests, it seems reasonable to assume that the dark matter which throws the calculation out is just that: ordinary, common or garden matter that is dark—stuff that doesn't give off light.

Even within our solar system, there is plenty of matter out there we can miss seeing, whether it's dust or bigger bodies that don't happen to catch the sunlight. In distant galaxies, it is only recently that we have been able to faintly (and often indirectly) detect the existence of a few planets and other bodies that don't shine as a star does. What's more, it is thought that many galaxies have supermassive black holes at their centers, with mass far beyond that of a star, and black holes are definitely dark. Isn't that enough? It's possible, but most cosmologists believe that dark matter is more exotic, a special kind of matter that doesn't act as normal atoms do.

MACHOS AND WIMPS

In reality it's an oversimplification that dark matter is considered to be a single substance. Broadly, cosmologists divide their ideas of what dark matter might be into two possibilities. There are some scientific names that just sound right, such as the photon. Others are downright weird, the gene "sonic hedgehog," for instance. But occasionally scientists try for humor and the result can be wince-making. Cosmologists divide up possible sources of dark matter into machos and wimps. A macho is a "MAssive Compact Halo Object." These are objects made of conventional matter, whether everyday such as a comet, or more extreme such as a black hole, that can't be seen because they don't emit enough radiation. They are thought to live in the galactic halo, the space around the galaxy.

Unfortunately there isn't much evidence for the machos providing anywhere near enough mass (we need enough dark matter to weigh a good seven times more than all the ordinary matter in the universe) so the alternative is something altogether more exotic, "Weakly Interacting Massive Particles." These can't be detected the way we would ordinary matter, probably not interacting with light at all, but are still heavy. We know of some lightweight particles, neutrinos that fit the "not interacting" part well, but they aren't heavy enough to add all that extra mass to the universe. There could, however, be something else out there—a kind of sumo-neutrino—that would add enough mass.

The existence of dark matter is given a theoretical boost by the results from the COBE and WMAP satellites, just as they also give insights into conditions soon after the Big Bang. The slight variations in cosmic background radiation are thought to be the outcome of quantum fluctuations that resulted in greater and lesser accumulations of matter. The slightly colder patches reflect the places where matter has scattered the radiation. But the amount of matter that can be deduced from these maps suggests there just wasn't enough ordinary matter to provide the gravitational impetus to form the galaxies we now see. There needed to be more attraction, presumed to be from dark matter.

Because the dark matter would not be blasted around by the still-violent radiation present when the universe started to become uneven (it does not interact with photons), it could condense more quickly to form the cores of the structures that would eventually cluster around it. If we had to wait for ordinary matter to overcome the battering of the radiation and cluster due to the force of gravity, and everything behaved as it does now, then there just wouldn't have been time for the structures in the universe to form.

Even dark matter as described thus far isn't quite enough to patch up the problems of structures forming. Although it's reasonably good at showing how galaxies may have come together, it doesn't explain the larger scale structure that seems to link groups of galaxies; for this, theorists have had to propose a second kind of dark matter (hot dark matter) which is more energetic, and thus less influenced by local gravitational attraction. Some see this as simply a necessary fix, and others yet another reason why the existing cosmological model is too creaky and patched up, over-ready for replacement.

DEFYING NEWTON

Dark matter isn't universally accepted. Some of the theories that compete with the Big Bang that we explore later have the added benefit of not requiring dark matter to exist at all. Alternatively, it could be something else that is causing that anomalous behavior of galaxies. Back in the 1980s, Mordehai Milgrom from the Weizmann Institute of Science at Rehovot, Israel, suggested that rather than hidden mass, it was our assumption that gravity made everything behave the same way, whatever its size, that was at fault.

Milgrom's idea, now generally known by the acronym MOND for "MOdified Newtonian Dynamics," suggests that gravitational forces act differently when dealing with something on the scale of a galaxy. We know some cases where operating on very different scales brings in altered behavior, the radically different way that a quantum particle and a macro object such as a baseball act, for instance, so why not a variation in the way bodies act under gravity when they get to a certain supermassive size?

Variants on MOND can cope with many of the observations that astronomers make which seem at odds without the existence of dark matter, but on the whole astrophysicists don't like theories that play around with physical behavior we believe we know well; they would prefer it if things were consistent across the universe. Once you throw away homogeneity and isotropy, for instance (see page 122), apart from anything else, the fundamental basis for developing the Big Bang theory falls apart.

DARK ENERGY

If dark matter is mysterious, it is at least a kind of something with which we're very familiar. Matter is the stuff of everyday life, of the Earth and the furniture around us, and our bodies. Energy is a rather more diffuse concept. Although we use the word "energy" all the time, it's in a very vague way. We tend to conceive of energy in terms of what it does for us, rather than as a separate distinct entity. And dark energy is even more unapproachable.

The existence of dark energy, or at least something that has the same effect as dark energy might have (remember this is all fairly untestable stuff), was first suspected when looking far back in time, peering into the distant depths of the universe. According to the Big Bang plus inflation model, after the inflationary period the universe was still expanding, but at a much more sedate rate. The assumption had always been that it was gradually slowing down, and at some point would grind to a halt and the whole process would reverse and gravity manage to pull the galaxies back toward each other.

This is a very sensible assumption. Gravitational attraction means that eventually the momentum that started the whole thing will run out and the contents of the universe will begin to move inward. The eventual outcome would then be a Big Crunch, the opposite of a Big Bang, as everything in the universe rammed together in a colossal collision.

In the 1990s, equipment was good enough to compare the rates of expansion way back, billions of years in the past, with the current expansion. By this time, they had better standard candles than in Hubble's day (see page 43), although the provisos about standard candles still apply. Instead of using variable stars, a particular type of supernova, formed when a white dwarf star consumes a companion star and becomes unsustainably large, was used. Because they are dependent on the size and type of star, such explosions are remarkably consistent in brightness, and therefore ideal for looking way back in cosmic history.

The teams studying the rate of expansion were in for a shock. The rate of expansion of the universe wasn't slowing down. It was increasing. Something was acting against gravity, actively pushing galaxies apart. This "something" was given the name dark energy. This is something we have seen cropping up already when Eddington referred to a "cosmic repulsion" needed to make the universe old enough to have been around to produce the stars and the Earth.

EINSTEIN'S GREATEST MISTAKE

Interestingly, dark energy was also something Einstein had once predicted could exist and then called his "greatest mistake." When theoreticians first began to apply Einstein's general relativity to the universe they found that there were solutions that could allow for a universe with expansion changing at different rates. If the universe was blasting apart fast enough, it would have effectively passed its own escape velocity and would never stop expanding. Gradually everything would thin out as it became bigger and bigger without end. The universe would sink into a "big freeze."

If, however, gravity was winning the battle, over time the expansion would become slower and slower, finally grinding to a halt and reversing, just as would a ball thrown up by hand on the surface of the Earth. In such a picture, as we have seen, everything in the universe would collapse back in on itself in the reverse of the

Big Bang, which has been referred to as the Big Crunch (or more idiosyncratically as the "gnab gib").

The middle solution had expansion of the universe that would go on forever, but didn't result in an infinite thinning out of the universe. This surprising possibility comes out of the nature of infinite series, which as we've already seen (page 33) can have an infinite set of components and still add up to a finite value.

Let's imagine that such a series describes the size of our expanding universe. It still expands. Let's say the first second it expanded one light-year, then the next second one-half a light-year, and so on. It would be expanding forever, but would never expand more than two light-years. You can blow up those values as much as you like: if the initial rate of expansion is right by comparison to the pull of gravity, it will never stop expanding, but it will never grow farther than a certain horizon.

These different types of universe correspond to different types of curvature of space. Einstein added a factor called a cosmological constant to his equations. This was before Hubble's observations suggested that the universe was indeed expanding. The cosmological constant (later called by the Greek letter lambda) was a fudge factor, a fix to force the universe not to expand out of control, because Einstein believed at the time that the universe was fixed in size, not expanding.

For a fair while lambda was dropped when it was found the universe really was expanding; this was why Einstein referred to it as a mistake. But the observations since the 1990s, showing an acceleration of expansion, require lambda to be in place working in the opposite direction to that originally suggested by Einstein. What was once a way to fix things to fit a theory has now become an active measure of a strange source of energy. Dark energy is lambda, the thing that is pushing the galaxies away from each other.

THE BIGGEST MYSTERY

Dark energy is no delicate phenomenon, only measurable with the most accurate of instruments. If it exists, dark energy is thought to amount to 70 percent of the energy and matter in the universe. (Remember that $E = mc^2$ means we can consider energy and matter interchangeable in terms of summing up the contents of the universe.) Just think of that for a moment. More than two-thirds of everything in the universe is this strange source of energy that is tearing the universe apart. It's more than twice as big as everything else added together.

Although there is no explanation of where dark energy comes from, there is a sensible explanation for why it appears to be acting the way it is. If you think of it as the energy associated with empty space (something that quantum theory tells us it should have), then the amount of this energy present is purely dependent on the amount of space available. As the universe expands, there is more space, and more of this energy. Eventually, the expansion is sufficient that the repulsive dark energy takes over as the main driving force. When we look back in time by using instruments to penetrate deep into space it seems that around 5 billion years ago, just before the Earth was formed, dark energy took over, generating the acceleration we have seen since, which has the universe doubling in size every 10 billion years.

This expansion doesn't sound like a lot stated that way but it is undergoing exponential growth. Exponential growth is just a change where the bigger the thing growing is, the faster it grows. The result is a graph of growth that starts slowly and then shoots up toward the end. Exponential growth doesn't just explode off

the graph, it rapidly escapes our comprehension. We find it difficult to anticipate the impact of exponential growth.

According to legend, the peasant who invented the game of
chess was offered a reward by a grateful emperor. (Another version
of the legend has an Indian king playing chess against a sage, who
turns out to be Krishna in disguise.) The peasant asked for what
seemed a trivial reward. One grain of rice for the first square of the
board, two for the second, four for the third and so on, doubling up
the amount for each of the sixty-four squares. The emperor agreed,
only to discover to his horror that before long he had used up every
bit of rice in his empire. To complete the sequence up to the sixty-
fourth square would require around 37,000,000,000,000,000,000
grains of rice. Admittedly on a long timescale, that's how the universe appears to be growing.

THEORY OR FACT?

Let's be absolutely clear here. Dark matter and dark energy are accepted parts of the current most widely accepted model of the
universe, but as was the ether before them, they are unlikely constructs, added to make an inexplicable behavior fit the natural
world. They are "place markers" in physics, not necessarily true
concepts. They may prove to exist, but it is entirely possible that
we will soon have an accepted model that disposes of them and de-
ploys totally different reasons for the behaviors that are explained
by dark matter and dark energy.

There is certainly already a range of theories that render them
unnecessary, including historical variation in the speed of light
and variations across the galaxy in the force of gravity, some of
which we cover in more detail later. For the moment, it is enough

to be aware that as with much of cosmology, these are not tried-and-tested phenomena that have been observed and measured, but rather absences of phenomena that are assumed to be caused by something unobserved and largely unobservable.

The real problem we have with being definitive about the Big Bang is that the only methods we have of looking back in time at the moment, based as they are on electromagnetic radiation, whether it's visible light, microwave, or radio, can't get past a certain point. If current theory is correct, the universe was opaque from the Big Bang until it cooled sufficiently to allow electromagnetic radiation free passage. So to see farther back requires something more subtle.

WAVING WITH GRAVITY

What is, without doubt, the most subtle telescope in existence at the moment is a linked pair of devices that go by the name LIGO (that's Laser Interferometer Gravitational wave Observatory). What makes this device subtle is that not only is it looking for something that's incredibly difficult to detect, it is searching for something that may not even exist. Now that *is* subtle.

At one time, gravity was regarded as something little short of magic. After all, it acts at a distance with no obvious connection between the two bodies that are attracted to each other. This seemed initially to be something similar to magnetism, which also acts with nothing visible connecting the magnetic poles. Now we know that magnetism's apparent action at a distance is caused by an interplay of photons, shifting energy between the objects. With Einstein's formulation of general relativity came the possibility that gravity works in a similar way.

It is now assumed that gravity is communicated at the speed of light by gravitons, the gravitational equivalent of photons (although gravitons have never been observed). It is also thought that there can be gravitational shock waves, coming out of the idea that gravity is not unlike stretching a rubbery membrane. It seems reasonable that sudden changes of gravitation in a region—for instance when a star explodes—should send out gravitational shock waves. Those same gravitational waves that we have already seen should be present from the early days after the Big Bang.

If these gravitational waves do exist, then we have a way to see back, through the optically opaque period, because gravitational waves would not be influenced by the limits on electromagnetic radiation. In principle we should be able to see the gravitational shock waves of the Big Bang itself. That's the kind of measurement LIGO sets out to make.

LIGO TO LISA

LIGO is based in two locations: Hanford, Washington, and Livingston, Louisiana, which are located around 3,000 kilometers apart. This spread gives them the opportunity to try to detect the origin of a source of gravitational waves, because they should arrive at slightly different times at the two locations. Each detector has an interferometer (in fact, one has two). These are a pair of tubes in an L shape, each around four kilometers long. The air is sucked out of the tubes and lasers shone down them.

If there is a gravitational wave, it should move the laser beams ever so slightly. To amplify this, the beams pass up and down the tubes seventy-five times before meeting. Because the two beams are at right angles to each other, they should experience a subtly different shift, and that means that the phase (the stage of progress on the wave, or a measurement of the state of the photons) will slightly shift with respect to the other, which will cause a shift in fringes where the two beams are brought together.

So far, so good. But there are serious problems here. First it is almost impossible to isolate the devices from extraneous vibration. They can't be used if anything sizeable is moving in the neighborhood. They are even influenced by waves crashing on the beach. So it is very difficult to isolate a shift that is definitely caused by gravitational waves. It is even harder to pin it down to something as definite but directionless as the Big Bang.

LIGO went into operation in 2002, ten years after the project was started. At the time of writing, all that has been shown is that some dramatic events out in the universe don't appear to have gravity waves associated with them. The LIGO team aren't too upset. They see LIGO as just the start; it isn't really sensitive enough to find the gravitational echo of the Big Bang, but the hope is to build a "son of LIGO" called Advanced LIGO, and a yet more advanced, and much more sensitive space-based version called LISA (Laser Interferometer Space Antenna).

LISA's great advantage, if it is ever built, will be that it can use much larger laser run lengths, around 5 million kilometers, and will be shielded from the vibration any earthbound device suffers. There are some doubts about its construction, however. Enthusiasm for big-spend science projects started to wane in 1993 when the Superconducting Super Collider was canceled. This huge project in Texas was already under construction when the plug was pulled.

To date this hasn't happened with Advanced LIGO, although the project has slipped significantly. Back in 2005, it was expected that LIGO II (as it was then known) would be in place in 2007, but current estimates put it back at least to 2013. LISA is currently only a proposal for a joint project between NASA and the European Space Agency. Expected back in 2005 to be launched in 2010, the earliest it is now envisaged launching is 2018. Arguably such research is much more valuable scientifically than a manned space program, however, there is no doubt the latter is where the publicity for NASA is, and hence where the budget is likely to be focused.

To be fair to those controlling the dollars, though, it ought to be emphasized again that LIGO is yet to come up with any detection of gravity waves. It is entirely possible that the whole concept is flawed, and any future investment would be pointless, but without the sensitivity of a LISA it is unlikely we will ever know for sure. If LISA could pick up the shock waves from the beginnings of the universe it should be able to distinguish between the kind of outcome expected from the Big Bang followed by inflation, compared with the different aftershock of the different theories we will soon come across.

It's hard not to have a real mixed feeling about a project such as LISA. Firstly, it could come to nothing at all. If Advanced LIGO, ten times more sensitive than the original LIGO, fails to detect any gravitational waves also, it might be time to rethink the theory, although arguably the sensitivity of LISA would be worth having for absolute confirmation. If we do get a spread of gravitational waves that can be pinned down to the very beginnings of the universe (not easy to be 100 percent certain) we still have to interpret those data, but in principle they would give some clues as to how the universe began. A lot of "if"s and "but"s certainly, but it does have the potential of providing a small window of understanding in a discipline that is mostly speculative.

Whatever the outcomes from LIGO and LISA in the future, for now the Big Bang is still accepted by most cosmologists, so before looking at other theories that give us a better potential to answer "what came before," let's temporarily assume it to be true, in order for us to be clear how time relates to the Big Bang and understand why with the most popular version of the theory, the concept of "before the Big Bang" simply doesn't exist.

8.

LET THERE BE TIME

Compare the length of a moment with the period of ten thousand
years; the first, however minuscule, does exist as a fraction of a second.
But that number of years, or any multiple of it that you may name,
cannot even be compared with a limitless extent of time,
the reason being that comparisons can be drawn between
finite things, but not between finite and infinite.

—ANICIUS MANLIUS SEVERINUS BOETHIUS (480–524)
The Consolation of Philosophy (trans. P. G. Walsh)

Before starting on this chapter, I'd recommend getting a cup
or glass of your favorite beverage, and sitting somewhere comfortable and quiet where the brain can operate freely. It's not that the
concept itself is a complicated one, but the implications tend to
twist the mind into pretzel form. Taking the lyrics of John
Lennon's "Imagine" even further, imagine there was no time, or at
least, imagine that time had a beginning at the Big Bang.

If it is truly the case that time started with the Big Bang, and
it's the suggestion most commonly associated with the Big Bang
plus inflation theory, then there is a very simple answer to what
came before the Big Bang. Nothing. Because there was no "before
the Big Bang." The four-dimensional model of the universe cre-

ated by Einstein and his mathematical guru Minkowski has the familiar three dimensions of space and one of time. What if those dimensions all had a clear starting point? What if, before the Big Bang, spacetime simply did not exist?

FOURTH-CENTURY TIME TRAVEL

To gain a basic insight into this idea, we're going to have to travel back in time, not all the way to the Big Bang itself, but to the fourth century AD. Because, incredibly, a man who was born back in 354 had some powerful insights into the nature of a universe where time and space came into being as a single entity. His name was Augustine.

He now tends to be referred to as Saint Augustine of Hippo (Hippo was a Roman city in North Africa, now Annaba in Algeria) to avoid confusion with the fifth-century Saint Augustine who founded the church in England and was the first Archbishop of Canterbury. Augustine of Hippo was one of the main figures in shaping the early Christian church, a "church father" who among other things came up with the doctrine of original sin.

Augustine was born of a farming family in Thagaste (now Souk-Ahras) in what is now Algeria. He didn't become a Christian until he was thirty-three, leading a worldly life that gave him insights that some of the other church fathers lacked. He famously commented that as a young man he had prayed, "Grant me chastity and continence, but not yet," as he was afraid that God might "too rapidly heal me of the disease of lust." Being a priest didn't come naturally to him—he had to be forced into ordination—but he was a natural writer, and this comes through strongly in his book, *Confessions*.

If Augustine had been alive today, *Confessions* would probably have been a blog. This sounds like a comment made to sound controversial, but I genuinely believe this is true. It has that same spontaneous feel, showing a personal nature and development over time, the kind of thing that typifies a good blog. Before plunging in to what Augustine has to say, we need to understand where *Confessions* came from.

It was written soon after he became a bishop in 396. His ordination caused considerable controversy, both because he was baptized abroad (in Milan, Italy) and also because he had tried out various other religions before settling on Christianity, notably the Manichean heresy that was considered such a danger at the time, and had attacked the Church's ideas. The criticism of Augustine was public and strident. His *Confessions* was his defense against his critics.

Why is *Confessions* of interest here? Because Augustine deals with the idea of time before creation, in effect, time before the Big Bang. Rather unexpectedly, he begins with a joke. Admittedly, he twists this around a little, saying that he won't answer the question, "What was God doing before he made heaven and earth?" the way someone else did, but nonetheless the joke is there in the alleged reply, "He was preparing hells for people who inquire into profundities." Augustine does then rather spoil the humor by saying that it wasn't a good idea to ridicule those who ask a deep question, but he put the joke in, in all probability to give a moment's lightness to the discussion.

Augustine's initial argument is simple. Assuming by "heaven and earth" we mean every created thing, rather than literally those two places, then God certainly wasn't making anything before the moment of creation because anything he made would be created, and you can't have created something before everything created was made.

Then, however, Augustine gets to the crux of the matter. He says, "If someone's mind is flitting and wandering over images of past times" and is astonished that God didn't bother to do anything for all the ages before the moment of creation, then that person should "wake up and take note that his surprise rests on a mistake." According to Augustine, just like Einstein, time was part of the same bundle as space. And if time didn't exist before that bundle of creation, there was no sense of God sitting around waiting for an arbitrary moment to begin creation. Before the creation there was simply God with neither time nor space. As Augustine puts it, there was no point asking what God was doing then because, "There was no 'then' when there was no time."

In the eternity that existed before creation, Augustine says, it was always the present. There was no past, no future. This isn't an easy concept to handle. Time, Augustine concedes, is a difficult subject on which to get a handle. "What, then, is time?" he asks. "Provided that no one asks me, I know. If I want to explain it to an inquirer, I do not know." It's easy to sympathize with him, yet it's important to grasp this idea of timelessness.

Once you have a feeling for the nature of timelessness, there is a simple answer to the question of why the Big Bang happened when it did. If there was no time and space "before" the Big Bang, then there can be no "when." In this picture, the Big Bang happened and time began. That had to be the start, because there was no before, thus there was no sense of something (it doesn't have to be God, it could be quantum perturbations if you prefer a godless universe) having to decide "now is the right time to do it." Something happened in that eternal "now" and that became the start of and definition of when.

There really is no need to go past Augustine in looking at the basic concept of a beginning of time coinciding with the beginning of space. We can't imagine ourselves outside of time, because our entire lives are conducted within it, but we can get a sense of that timelessness if we take a look at what's meant by past, present, and future. (Augustine also did this; I am borrowing his concept, but presenting it slightly differently.)

PAST, PRESENT, AND FUTURE

Does anything other than the present exist? Arguably not. The past is fixed and can't be influenced. We can't see it or touch it. We are dependent on fragile (and often flawed) memory to give us a connection to the past. Take a simple example. A little while ago I got an e-mail from a friend, saying he'd seen me out walking my dog while talking on a cell phone. I knew this wasn't true, partly because I hadn't taken the dog out that afternoon and also because I never take a phone with me when I walk the dog, as it spoils the opportunity to think.

Now just imagine, just to really emphasize the point, that he saw "me" go on to commit a murder. According to my friend's view of the past, I committed a murder. He saw it with his own eyes, according to his memory. Yet according to my view of the past (the correct one in this case), I know very well I wasn't even there. A court of law could take the version of events his way and find me guilty. That would become the official version of the past, but it wasn't what truly happened. The past is little more than chains of memory.

Some have argued that this is an outdated view. Memory was our only link to the past before technology took over. We can now rely on photographs and video to bring the past alive. If we're honest, many of our "memories" of earlier events aren't true memories at all now: they are the images captured by cameras. However, this argument misses the point. Even if you have all those photographs and videos, you aren't looking at them all right now. In the now, in this moment, even your photographs and videos are only a point of reference in your memory.

As for the future, we're not dealing with something that is fixed, but rather a collection of probabilities. Think about tomorrow. I might die in the night. (Apologies for the morbid streak, but it helps to make the point.) For me, there would be no tomorrow in the way I normally anticipate tomorrow. I can't say that tomorrow is in any sense real. The Greek philosopher Aristotle came up with a useful analogy for thinking about the future. He suggested we think about the Olympic Games.

Aristotle suggested there was a different class of thing to the everyday, a potential thing. Do the Olympic Games exist? Certainly, we can't doubt that. But imagine an alien appeared in a flying saucer (that bit's mine rather than Aristotle's) and asked you to show it this "Olympic Games" thing. You couldn't. It does exist, but not in any sense that you can currently touch it and feel it (unless, by a huge coincidence, you are reading this book while sitting at an Olympiad). The same goes for the future. It has the potential to be, but it can't be experienced.

Therefore the only reality is the infinitesimal flick of time that marks the present. The now. We think we can deal with longer and shorter periods of time, but in reality what we are talking about is either the memory of time past or the anticipation of time to come. We don't experience long times. Imagine you had to wait eight hours to see someone in a featureless waiting room without a magazine to read or anything to do. You would say that this was an agonizingly long wait. But really you would not experience a long time.

You would anticipate what a long wait was ahead of you. You would remember a long wait after you'd done it. But you would only ever experience the fragmentary moment of now.

If we take an Augustinian view of the Big Bang (in itself a strange distortion of time, as we are using the viewpoint of a man who lived 1,600 years before the name was even dreamt up), then we begin with the "now" of the Bang itself. No past to worry about. No need for a causal push, which inherently implies a "before." Just the Big Bang in a "now" that resembles the now in which you are living. Now, this moment. As you read this next word. Now.

I really do recommend taking a moment's pause (itself a fragile concept in this timeless view of consciousness) to get your mind around these ideas. Play around with the idea of past, present, and future in your mind. Get a feeling for the concept of time beginning, having no "before" its creation.

The idea that time began with what Augustine would call the creation and scientists now call the Big Bang is not just a matter of the opinion of a fourth-century cleric still being referred back to today (although I do find it fascinating that almost every book on the Big Bang does mention Augustine, and it is well worth taking a look at his *Confessions* even now). It is a real conclusion that comes out of our modern understanding of space and time that happens to align with Augustine's ideas, so he still gets dragged in.

SPACE, TIME, AND RELATIVITY

The reason it is thought that time began with the Big Bang is because this wasn't the emergence of the universe into space, but rather the emergence of space itself. Before the Big Bang, the standard theory assumes, there was no space: not empty space, just nothing. And the Einsteinian picture of the universe merges space and time into a single entity: no space, no time, because the whole thing is one: spacetime.

It's easy to think that this idea of "time as the fourth dimension" and linking time to space to make this unified concept of spacetime is just a bit of terminology. You might suppose that what the scientists mean is that time is "a bit like" a spatial dimension. But Einstein showed that time has a fundamental, unbreakable link to space. Relative motion in space results in real changes in the nature of the passage of time.

Take the simple case of a spaceship, heading away from the Earth at half the speed of light. By the time it is 10 light-years away (a journey that we will see as taking 20 years) the clocks on board will be running 5¾ years slow from our point of view. Say this ship left the Earth in 2020. Then if we could somehow take a instantaneous look at that ship in 2040, on the ship it would be 2034. This is not just a strange visual effect, in the way that perspective seems to make things look smaller than they are; the time on the ship really would have run that much slower from our point of view.

Special relativity, which deals with objects that aren't accelerating (thus are moving in a straight line, as acceleration simply means change of velocity, and turning is a change of velocity because velocity has both speed and direction components), is joined by general relativity, which deals with acceleration and gravity. From general relativity we discover that a gravitational field alters time too.

On an all too regular basis as editor of the www.popularscience .co.uk book review site I am sent self-published books that set out to prove Einstein wrong. For some reason, he seems to be a very popular target for those who have developed their own flaky theories and want to take on the world of science. But in reality, we are yet to see any contradiction to relativity stand up to serious consideration. Experimentally, this linkage of time and space has been shown again and again.

According to legend, when the Global Positioning System (GPS) satellites that are used for satellite navigation in cars and planes were first planned, the military (who own GPS) wouldn't believe that it was necessary to allow for such a theoretical concept as relativity in designing navigation satellites. But the fact is, it was essential. Both special relativity and general relativity influence the flow of time on those satellites as seen from the Earth. Special relativity slows them down and general relativity speeds them up, but the two effects don't cancel out, so GPS satellites have to make a correction to deal with the relativistic effect.

As the way GPS works depends on comparing a time signal from each of several different satellites, each carrying its own atomic clock, this change in time caused by relativity had to be allowed for, or the system would drift out of synchronization. The effect is real and has an impact on our everyday lives. So there should be no doubt that time and space are linked rather than entirely separate entities. If spacetime truly all started at the beginning, there is no worrying about why the Big Bang happened at a particular point in time. It didn't. It just happened, and from then on there was time.

We can't really look for positive evidence for a big nothing in time and space before the Big Bang, only for ways to disprove it, but this is not a problem, because this is how most scientific methodology works. As we have already seen, it isn't possible to absolutely prove something; we can only examine the best evidence to support a theory. As yet there is no particularly good evidence that spacetime began with the Big Bang, but as more data are gathered we will either find that there is nothing to disprove this concept, which will make it increasingly attractive, or that some data contradict the theory and destroy it.

Either way, it should not be forgotten that there might have been no "before" as we delve into the alternative theories of what came before the beginning. Some of these theories dispose of the Big Bang altogether, but that isn't necessary. As we will see in the next chapter, even if the Big Bang was the origin of our current universe, it is entirely possible that time and space did not begin around 14 billion years ago.

9.

GROUNDHOG UNIVERSE

> After all, the "universe" is an hypothesis, like the atom, and must be
> allowed the freedom to have properties and to do things which would
> be contradictory and impossible for a finite material structure.
>
> —WILLEM DE SITTER (1872–1934),
> *Kosmos*

In the movie *Groundhog Day*, Bill Murray's character Phil Connors lives the whole day over and over again. In a much larger-scale version of the *Groundhog Day* scenario, some cosmologists believe that the universe cycles through endless collapses into a Big Crunch followed by endless rebirths in new Big Bangs. This would mean that before the Big Bang there was another universe and before that another, potentially in an endless sequence without limit back in time. With enough repeats, in principle every possible life could unfold as time after time a universe would come into being, pass through its many billion years of life, and be destroyed.

Such a cyclic universe can be pictured by thinking of the way a gasoline engine operates. Because it's cyclic, it's arbitrary where we start, but let's begin with the Big Bang. In the gas engine there's an electric spark that ignites the fuel, driving the piston so the volume contained in the cylinder expands. Eventually the piston reaches its extreme point and the volume in the cylinder begins to contract, until with practically no space left, another spark sets the whole process in motion again. The *Groundhog Day* universe, like the gas in the cylinder, expands and contracts time after time, powered by the spark of a series of Big Bangs.

CYCLING THROUGH THE UNIVERSE

There's something very satisfying about such a cyclic model. There is no need for all the complex arguments about inflation, because the new universe after the Crunch and Bang would be based on the previous one. If that earlier universe were already uniform and flat and had galaxies, its new offspring would be seeded by the earlier structures. It could carry through the uniformity and flatness without the need to create them from scratch, as is required in the Big Bang.

This model requires whatever it is that's pushing the universe apart quicker and quicker at the moment—dark energy—to decay over time, so that eventually gravity reasserts itself and the universe begins to contract. But this is an assumption that is certainly no more far-fetched than those that are piled on top of each other to make the inflationary Big Bang work.

In a cyclic universe there is a very simple answer to the question, What came before the Big Bang? First of all, there was no true Big Bang in the sense of an origin of time and space. But before the Big Bounce or Big Crunch and Bang, this theory envisages a very similar universe, the structure of which had a direct impact on the way our own universe was put together. It was Robert Dicke, who was central to the early work on the cosmic microwave background radiation (see page 141), who first strongly promoted the cyclic or oscillating universe. He was unhappy with the shoulder shrugging that seemed to be the only response to the question of why the Big Bang should ever have occurred in the first place.

UNIVERSAL GARBAGE DISPOSAL

Dicke wanted a picture that didn't just start at an arbitrary point in time (or with time), but that had an explanation for why that Bang had happened, and what came before. Dicke's picture of a universe going through a repeated cycle was neat, although it generated some problems of its own. One was of garbage disposal. George Gamow had been proved wrong when he assumed all the chemical elements were made in the Big Bang. It is now clear that only the lightest elements were there at the beginning, and the rest were produced in stars and supernovae.

This didn't present a problem for those dealing with a simple Big Bang picture, but it did introduce an issue Dicke would have to explain. Before one of the Big Crunches, the previous universe would have all the heavy atoms present, just as our cosmos does today. Where did they go as the universe passed from Crunch to Bang? How come those heavy elements weren't there immediately after that initial bounce?

Dicke decided that the temperature and pressure in the Crunch were sufficiently drastic to smash the heavy atoms into their component pieces. It takes a lot to do this. Remember, stars don't break atoms down; they build them up. It would take much more violent conditions than those in a star to shatter those heavy elements to their constituent parts. But Dicke believed that the conditions at the Crunch/Bang interface would be so hellish that this atomic disassembly would take place.

And dealing with that was not the only problem that supporters of a cyclic universe had. Not only are there still issues with the singularity at the Big Bang, there is a problem that is reminiscent of the gasoline engines mentioned at the start of the chapter. There is no such thing as a totally efficient engine; there are no perpetual motion machines. This falls out of the second law of thermodynamics (see page 273). Yet for a cyclic universe to carry on forever would seem to imply that either it is a perpetual motion machine or that it will eventually run out of energy and give up.

NO PERPETUAL MOTION

With further analysis, a simple *Groundhog Day* universe was shown not to be practical by looking backward in time. As the universe expands, the amount of radiation present increases. This would then be squashed together in the Big Crunch, so there's more radiation present at the start of the next cycle. The more radiation the longer it is before the expansion runs out of steam and the contraction begins. So a later cycle in the sequence would be longer than an earlier one.

All you then have to do is extrapolate backward. If later cycles are longer than earlier ones, the cycle before ours would be shorter than ours. The one before that would be shorter still. If we add those lifetimes together, it's possible that it converges on a time in the past, before which there would be no cycle because it's of zero length, and in that case, the universe had a beginning, so we're back to the same issues as many have with a Big Bang starting for no obvious reason.

Although this seems a self-evident thing—that a decreasing length of time for each previous cycle means that there was an identifiable start time for the universe—it is not necessarily true. Admittedly if, for example, each cycle were twice the length of the previous one there would be a clear start. Imagine, for example, our current universe had a lifetime of 1 unit. Then the previous one would be ½ unit, the one before ¼ unit, and so on. Thus the entire life of the cyclic universe would be $1 + ½ + ¼ + ⅛ + 1/16$. . . . You are back to the sort of finite sum series we have already met, which could fit an infinite set of cycles in just twice the lifetime of the current universe.

If, however, the previous universe lasted ½ the current one, the one before ⅓, the one before ¼, and so on, then the entire lifetime of the cyclic universe would be $1 + ½ + ⅓ + ¼ + ⅕ + ⅙$. . . . Unlike the previous series, this one doesn't add up to a finite number. With an infinite set of entries you get a total that goes off to infinity. A universe that increased in age in each cycle in this way would have been around forever. But the analysis done on lifetimes by California Institute of Technology professor Richard Tolman in the 1930s showed that a simple cyclic universe would have a converging history: there would have been a finite time in the past when the cycles first started.

Even worse, when by the 1950s it was possible to better analyze how a contracting universe would behave as it approached the Big Crunch it was found that it would shift into a very unstable state where the size of the universe was fluctuating wildly as it headed toward the Crunch, resulting in the very opposite of the observed state: a universe that was anything but even and homogeneous, with huge variations in density, a start in life that would not produce a universe like our own.

FROM THE UNIVERSE TO FORCES AND PARTICLES

This doesn't mean that a cyclic *Groundhog Day* universe is impossible. A theory has been developed in the last few years that really could work, and that removes all the issues that the original Big Bang theory had. To see how this very different picture works, it's necessary first to look away briefly from the distant parts of the galaxy and get much closer to home. Modern theories of a cyclic universe have emerged from thinking about the most basic aspects of existence. Scientists have long sought a "theory of everything" that would explain how all the forces and particles that make up our universe interact. This was something that obsessed Einstein in the later years of his life, which he spent in a fruitless attempt to pin down an all-encompassing theory.

We know of four fundamental forces that describe how matter works. The weakest two are the most familiar. These are gravity and electromagnetism, the latter responsible for everything from powering an electric motor to the existence of light. The other two are nuclear forces, the weak force and the strong force. The weak is a phenomenon that is responsible for some aspects of nuclear decay, such as beta decay when the nucleus throws out an electron. The strong force has the job of holding the nucleus of an atom together, despite positively charged protons all repelling each other electromagnetically.

It seems overly complicated, having so many forces. Over the years there have been many attempts—Einstein's included—to combine them into one. There has been some success. Electromagnetism and the weak force proved easiest to pull together, and the strong force was also eventually combined with these, but then there came what seemed an insuperable barrier. Gravity is so different from the other forces that it has proved impossible to come up with a practical overarching theory that would handle everything.

I have mentioned Einstein several times already in this chapter, and there's a good reason for that. He provides a kind of interface between the two incompatible parts of our understanding of the universe. It was Einstein who developed general relativity, which describes how gravity works, and Einstein who was in part responsible for the development of quantum theory, the science of very small things on the scale of atoms or smaller, even though he hated the probabilistic nature of its approach. For all its subtlety, general relativity is a "classical" theory: gravity is a force that behaves in the way Newton could understand. The other three forces are quantum interactions.

Thus the way we understand these two different aspects of the universe is incompatible. Quantum theory works on the very small and provides us with three of the four forces. Gravity takes over as the scale grows to deal with the kind of objects we are familiar with in the human world. It's Einstein's bogeyman, quantum theory, versus his brainchild, relativity.

Ironically, this clash of the titanic forces behind existence comes to a head in the idea of the Big Bang. Because the Big Bang is a phenomenon of the very small, the starting point of the universe should be one where quantum theory dominates. In fact, it has been speculated that initially the different forces we now know did not exist. In the standard picture of the Big Bang, under the immense temperatures and pressures immediately after the Big Bang, all the forces except gravity would be melded into a single superforce, that "came apart" in a process known as symmetry breaking as that very early universe cooled.

However, that still leaves gravity separate, something that

doesn't seem possible if the initial universe were on a quantum scale. Many scientists would say that Einstein's mental conflict has now been resolved, thanks to an approach called string theory and its big sister, M theory. M theory makes it possible to pull gravity into line with the other forces as part of that initial superforce, and this theory is, without doubt, very neat, as it provides a simplification to a problem that by the 1970s was proving a real headache for physicists.

THE PARTICLE ZOO

Ever since ancient Greek times, it has seemed likely that matter is composed of small basic building blocks. Although not all Greeks believed it, preferring an alternative theory that had everything constructed of earth, air, fire, and water, some thought that matter was made of tiny component parts. The idea was you would take something, a piece of cheese, for instance, and cut it smaller and smaller and smaller. Eventually you would have a piece that was so small, so basic that with the sharpest knife it would no longer be possible to cut it. Such a piece was *a-tomos*, un-cuttable. An atom.

In that ancient Greek view, each type of material had a separate atom. Cheese atoms were different from plant atoms, which were different from metal atoms, and so on. They even thought the various atoms would have different shapes and colors. This atomic theory was largely ignored because the earth, air, fire, and water theory triumphed. It was nearly 2,000 years before atoms would come to the fore again. In their new incarnation, atoms supported the idea that everything was made of elements, chemical components that were impossible to break down further, whether they were gases such as hydrogen and oxygen, or solids such as carbon and iron.

By the twentieth century we had a true atomic theory, showing how every atom was made up of just three types of fundamental particle: neutrons and positively charged protons in the nucleus, with electrons in a statistical cloud around the outside. It seemed that we could deal with a nice simple set of particles. But then other particles were discovered. Some were in cosmic rays, crashing into the atmosphere from the stars. Others were produced in experiments where atoms were smashed into each other.

When I was doing my physics degree at Cambridge in the 1970s, it seemed as if practically every week a lecturer would rush in and announce with excitement that a new "fundamental" particle had been discovered. What had been a nice neat structure of three basic building blocks became a complex mess, not helped by the discovery that forces are also carried by particles: electromagnetism, for instance, by photons of light.

The good news is that since then a number of the particles have been pulled together with the discovery that many of the heavy particles such as neutrons and protons can be built from more fundamental particles called quarks. (The word should rhyme with "dork" not with "bark," but is rarely pronounced that way.) Even so, there is still quite a zoo of particles in the standard model. Physicists have always dreamed of a "theory of everything" that would pull together all the diverse forces and particles in a single theory that works neatly together. Supporters of string theory say that theirs is the first usable theory of everything.

HOW LONG IS A PIECE OF STRING?

String theory is so neat, because it does away with that zoo. According to string theory, every particle in matter, whether it's an electron or a quark, and every force particle from photon to graviton, is made from the same fundamental building block, a string. These incredibly tiny strings exist in closed loops (although there can also be open strings), and vibrate in different fashions, the nature of the vibration producing the different kinds of particle.

The nice thing about this model is that getting the hang of it is very simple. Anyone who has played a bowed string instrument such as a violin will know that by bowing the string in different places it is possible to produce harmonics, higher notes where the string vibrates in a different pattern. String theory works the same way. You could imagine a loop of string where the whole thing is vibrating as half a wavelength (the distance from one peak to the next on a wave) or a full wavelength, or one and a half wavelengths, and so on. The picture string theory gives us is that each of these harmonic-like variations results in a different particle.

It ought to be stressed that the strings in string theory aren't literally pieces of string. Pretty obviously they aren't made of string (which itself is composed of atoms) but also they aren't truly strings at all. The string description is a model. Just as we can think of light as being a particle or a wave, but it isn't truly either of these—it's just light—so similarly we can think of particles as being built up of strings, but really we mean abstract constructs that happen to have behavior that reminds us of vibrating strings.

This theory is pretty well unique in the history of science because the theory dropped out of the math, rather than the other way round. The usual approach would be that someone would observe the real world and would build a theory around what was observed. With string theory, there was some interesting abstract multidimensional mathematics, which happened to have some behaviors similar to the real world. This was then used as the basis for the theory. String theory also has the small problem of requiring ten dimensions to operate in, of which more later. Even so, it is a powerful theory which now has hundreds of scientists working on it.

USE YOUR BRANES

M theory, which will have particular significance for what might have come before the Big Bang, is an advance on string theory which adds in an extra dimension to the ten its predecessor requires, making ten spatial dimensions plus time. M theory has as a basic unit a "brane" which is a multidimensional membrane. This can have as many dimensions as you like up to the ten spatial dimensions, but in a one-dimensional form twisted through various other dimensions simplified to a string (making string theory a subset of M theory).

M theory describes our universe as a three-dimensional brane, floating through higher dimensions of space. Its development proved something of a relief to many string theorists, as it unifies five different, incompatible versions of string theory that had cropped up before then by adding that extra dimension. One of the strangest things about M theory is that no one seems to know why it is called that. The "M" has been said to stand for membrane (which makes sense) or mystery or magic; it's rather strange that the physicist Ed Witten from the Institute of Advanced Studies at Princeton, who first came up with M theory, has never clearly explained why it's called that.

As we will see, the existence of M theory makes for a dramatic alternative view of how our universe came into being, but it is also the case that a number of respected scientists don't consider string theory and M theory to be science at all.

Putting these concerns aside briefly, M theory does offer a

highly dramatic alternative view of how universes come into being. To begin with, to get a feeling for what M theory proposes,
you have to get your head around a universe with more than four
dimensions. We are used to the three everyday physical dimensions, plus time, but M theory demands we bolt on many more dimensions, bringing the total to eleven.

This is impossible to envision directly, so don't try. Going beyond
three physical dimensions has been a mainstay of both science fiction and fantasy writers and mathematicians for years. In fiction "the
fourth dimension" or a "parallel dimension" was a 1950s mainstay
for an alternative world operating alongside our own; it was more
used to describe another universe than literally another dimension.
In math, dimensions are just rows of numbers. Two dimensions are a
table like a spreadsheet. Three dimensions are a stack of tables, like a
series of parallel spreadsheets. Four dimensions are a table where
each element is a three-dimensional table, and so on as far as you
like. As far as mathematicians are concerned, it's as easy to have a billion dimensions as it is to have three (although there are a lot more
numbers involved).

Thinking physically, each of the three physical dimensions is at
right angles to the others. A fourth physical dimension would have
to be at right angles to all three of those. We can't see that fourth
dimension in our 3D visualization of space. If something moved
in the fourth dimension it would disappear from one spot and
reappear on another when it moved back to the same point in the
fourth dimension. You can envisage this happening by imagining
briefly what the equivalent would be in a two-dimensional world.

Imagine you were in the pages of a comic book, a character in
one of the illustrations. You could see another of the characters
moving from side to side or up and down. But imagine someone
in the three-dimensional world lifted one of the characters out of

your frame of the comic and dropped them down in another. He would simply disappear from one place and appear in the other. Until he reached the same slice through the third dimension that you occupy, the page, he wouldn't exist in your comic book world.

Similarly, if something were moved through a fourth physical dimension in our universe it would simply disappear, and stay out of our view of existence until it came back into exactly the same position in the fourth dimension that our universe occupied. In effect, a fourth dimension takes us from having a single universe to having an infinite series of three-dimensional universes, each at one of the points along the fourth dimension.

That's just adding one dimension. To make the math of string theory, the predecessor to M theory, work it was necessary to move on up from four dimensions to ten, whereas M theory adds yet another dimension as we'll see when we get a better idea of what this theory is. This implies there could be an infinite set of parallel universes, each shifted infinitesimally along some combination of those extra seven dimensions.

CURLED-UP DIMENSIONS

As we have seen, string theory, the predecessor of M theory, imagines that every particle is composed of a tiny one-dimensional "string" that vibrates in different ways in these various dimensions depending on what kind of particle it is. In effect, the extra dimensions are required to enable the different types of vibration necessary, but are described as being curled up very small so that we can't detect them.

This, when you come to think about it, is a strange concept.

Many popular explanations of string theory say that these dimensions have to be curled up very small, smaller than an atom, or things would always be drifting off into the extra dimensions and would disappear. Yet it's reasonable to ask why it was ever envisaged that such dimensions would be directly detectable. Go back to a comic book world. We don't have to say that the characters in the two-dimensional universe of the page are faced with a third dimension that is smaller than an atom just because they don't escape into the third dimension.

This seems to be a case of oversimplification on the part of the scientists coming up with the explanation. They don't really think the extra dimensions are curled up small because we can't experience them in everyday life; it's because they have to be like that for the math to work out. Remember, this is a theory where the mathematics came first and theory was fitted to it, rather than developing a model that reflects observed reality.

String theory first emerged from an accidental observation that an abstract mathematical equation first used in the nineteenth century happened to reflect very closely what was observed in a particle interaction. Building from this strangely back-to-front origin, string theory would be built up and up until it was a mathematical abstraction that in principle provided a working theory of everything.

From a technical viewpoint, string and M theory are very attractive. Not only do they seem to offer a solution that spans all the requirements of dealing with the fundamental particles and forces of the universe, they overcome one of the principal difficulties of the traditional view of particles. An electron, for example, is usually considered a point particle, infinitesimally small. This means that in principle the forces it produces should head off to infinity as we get closer to it.

Mathematicians can't cope with infinities as values in an equation, so this has to be fudged around. When these infinite values first emerged in the theory of the interaction of quantum particles, the response was horror. A good example is quantum electrodynamics (QED), the science of the interaction of light and matter. This has been described as the best theory in all of physics, in that it predicts most exactly what happens in reality. However, it did originally produce some infinite values and the developers of the theory had to find ways to make the infinities go away while keeping the theory. They did achieve this, but always regarded the fix with some suspicion. Strings, although incredibly small, aren't points, so they don't suffer from this problem.

TRUTH AND CAREERS

But there is a major concern about both string theory and the more recent M theory. There has been a huge amount of effort exerted on them since 1968, when the first glimmerings of string theory emerged. Many physicists have spent their whole careers working on it, and some would argue that this is why a theory that should have been thrown out long ago still has currency: there is too much invested in it.

This attitude of "It has to be true, I've given my career to it!" is shown in a comment from string theorist and science popularizer Michio Kaku: "If string theory itself is wrong, then millions of hours, thousands of papers, hundreds of conferences and scores of books (mine included) will have been in vain." As physicist Lee Smolin points out, referring to academia at large, "String theory now has such a dominant position in the academy that it is practically career suicide for young theoretical physicists not to join the field."

The problem with string theory and M theory for some is that it isn't what they would describe as a true scientific theory at all but rather a mathematical dead-end. They point out that it doesn't provide a clear identifiable solution, that what is observed in our universe is just one of an uncountable number of solutions the theory throws up. String theory gives no reason to choose that particular solution. Where quantum electrodynamics makes startlingly accurate predictions, but has trouble with infinities, string theory is fine with infinities but makes no new predictions. Because of this, it is arguably useless as a scientific theory.

This isn't just the opinion of nontechnical commentators. Lee Smolin is one of the most respected physicists alive, born in New York City in 1955 and now a researcher at the Perimeter Institute for Theoretical Physics, which he helped to found, and a professor at the University of Waterloo in Ontario, Canada. He has worked for several years on string theory and said, "Much effort has been put into string theory in the last twenty years, but we still do not know whether it is true. Even after all this work, the theory makes no new predictions that are testable by current—or even currently conceivable—experiments. The few clean predictions it does make have already been made by other well-accepted theories." Smolin also quotes Nobel Prize–winning particle physicist Gerard t'Hooft as saying he wouldn't call string theory a theory, or even a model, just a hunch.

Then there's the late Richard Feynman, arguably the greatest U.S. scientist ever, who strongly disliked the way string theory (known as superstring theory in his day) made arbitrary decisions to match reality. String theory, for instance, assumes that six of the ten dimensions are wrapped up small, but it gives no explanation as to why six should be curled up rather than, say, seven. "In other words," Feynman said, there's no reason whatsoever in superstring theory that it isn't eight of the ten dimensions that get wrapped up and that the result is only two dimensions, which would be completely in disagreement with experience. So the fact that it might disagree with experience is very tenuous, it doesn't produce anything; it has to be excused most of the time. It doesn't look right.

Although vast amounts of work have been done on string theory since, Feynman's concerns still apply. If it seems I am laboring this point, it's very important to understand that there are more professors and graduate students working on string theory and M theory than any other approach to dealing with the basic workings of physics. Vastly more. In that sense, this is the best accepted theory, yet it is still hugely controversial. However, if M theory is right it does give us a very strong alternative to the conventional picture of the Big Bang and what came before it.

It doesn't help that the math required to deal with string theory and M theory is so complex that most scientists haven't a clue whether it makes any sense. Here's the noted physicist Paul Davies on the matter:

It leaves string/M theorists without much of a reality check. Where this enterprise will end is anybody's guess. Maybe string/M theorists really have stumbled upon the Holy Grail of science, in which case one day they might be able to tell the rest of us how it works. Or maybe they are all away in Never-Never Land.

In principle there *are* some potential experiments that could support string theory. For example, string theory predicts that every particle we know has a "superpartner," a much more massive equivalent particle with a special relationship with the particle we know. As yet none of these have been observed, but it is possible that the Large Hadron Collider (see page 243) at CERN will produce them. String theory also makes it possible for particles to have weird and wonderful fractional charges, say $\frac{1}{3}$ of the charge of an electron, which again has not been observed. Yet this kind of evidence is pretty indirect, and even if it had been observed (which, bear in mind, it hasn't) it would only fail to disprove string theory; it isn't the kind of evidence to decide between theories that is generally looked for. As Lee Smolin suggests, there are other theories that such evidence could equally support.

Whether the solid math of string theory and M theory will ever prove useful is one issue, but there is no doubt that the conceptual descriptions of the universe provided by string theory and M theory enable us to envision very different kinds of universes with unique ways of coming into being.

INSIDE THE BLACK HOLE

String theory makes possible the bizarre concept that our entire universe resides in a black hole. We have referred to black holes, stars where none of the light escapes, a number of times already without really establishing just what they are. It's worth taking a moment to clarify this; most of us probably have a vague idea of what a black hole is, but that idea may well be formed more from fiction than physics.

The starting point has to be a warning that black holes are theoretical constructs. No one has seen one directly. They've certainly not been experimented on. Instead, as with our investigation of the Big Bang, we are left with theory and indirect observation.

There are minority theories that would explain everything we ascribe to black holes without the real thing existing, but to be fair to black hole supporters (and that's the vast majority of astronomers and cosmologists) there is a much more definitive theoretical basis for their existence than there is for many other cosmological phenomena.

The reasoning behind our assumption that black holes exist is twofold. They seem to be an inevitable conclusion of certain physical processes (although those processes don't have to have ever happened in reality) and various observations seem to suggest black holes exist.

The theoretical reasoning is down to observations of how stars change and develop through their lifetime. We haven't been around long enough to watch a single star go through this process, but we've observed enough stars in different stages of their development to make it very likely that these theories are correct.

Remarkably, black stars have been the subject of speculation for over two hundred years. John Michell, an astronomer and geologist from England born in 1724, who like so many other of his fellow British astronomers ended up working at Cambridge, was thinking about the concept of escape velocity, something that eventually would be an essential consideration for the space program. If you throw a ball in the air, it falls back down to Earth. It can't get high enough before the downward acceleration from the Earth's gravity slows it to a stop and sends it falling back. Superman, however, can supposedly throw a ball as fast as 11.2 kilometers per second (around 25,000 miles per hour), which means it escapes before gravity drags it back.

It might seem that this limit makes it impossible to send a rocket into space. Anyone who has seen a launch from Cape Canaveral will know that they take off much slower than 25,000 miles an hour, but escaping the Earth's pull is easier than it sounds. First, we can cheat a little by sending something off into space eastward, against the Earth's rotation, and near the equator, which means we have to achieve only around 10.7 kilometers per second. But more important, the farther away from the Earth you are, the lower the escape velocity is. Because a rocket is constantly under power it can take off relatively slowly, getting to a height where the escape velocity is lower and lower. If Superman throws a ball into space it has to start off with that escape velocity, because nothing is pushing the ball upward once it leaves his hand.

Michell imagined the escape velocity required to get away from bigger and bigger bodies. As the mass of the planet or star gets bigger, then the escape velocity increases too. What would happen, Michell wondered, if the mass got so great that the escape velocity was faster than the speed of light? Then light would never make it away from the star: no light would get out. It would appear to be a black star even though it was blazing furiously on its surface. (Michell didn't call his special star a black hole; this name was given to them by American physicist John Wheeler as recently as 1967.)

No one took Michell's idea, published in the *Philosophical Transactions of the Royal Society* in 1783, forward, and it wasn't until the early part of the twentieth century that anyone would come up with a way to envisage black stars with mathematical precision. Einstein's newly minted general theory of relativity predicted that light would be influenced by gravity, as the gravitational field of a body bent space, sending the straight light beam around a corner.

In 1916, while fighting in the First World War, German physicist Karl Schwarzschild found a way to describe the action of a star on light mathematically, using Einstein's equations. Of itself this was no surprise (except perhaps Schwarzschild's ability to under-take serious mathematical work on the battlefield) but a strange possibility dropped out of the math. Just as Michell had found with his basic assumptions on escape velocity, Schwarzschild showed that a massive enough star would bend space so far that light would never get away.

He thought this was only a mathematical nicety with no real application, because the ability to bend space was dependent on both the mass of the star and its size. It wasn't enough to have a supermassive star, it would also have to be much tinier than was the case with any star that had been observed. To get our Sun, a small to middling star at 1.4 million kilometers in diameter, into a state where its mass were concentrated enough to go black, you would have to compress it smaller and smaller until it was only around 3 kilometers across.

However, with further work from Indian physicist Subrah-manyan Chandrasekhar and American Robert Oppenheimer, it became apparent there was a way for such compacting to happen. Any star has a huge amount of mass, and all that material in the star is pulling together with gravitational attraction. While the star is very active, the outward pressure from the nuclear reactions that power it keeps the star "fluffed up," but as the nuclear fuel runs low, that pressure will drop and the star will begin to collapse.

Now another physical force comes into play, a quantum fea-ture called the Pauli exclusion principle which means that similar particles of matter that are close in distance must be different in velocity. This will counter the gravitational collapse as a star cools unless the star is so massive that gravity overwhelms. The mass re-quired for this to happen is around one and a half times that of the Sun. Some such stars explode as a supernova, seeding the universe with heavy atoms. But if this fails to happen, the star should contract, getting smaller and smaller until the gravitational intensity is such that space is so curved that light never escapes: it has become a black hole. In theory, though, there is nothing to stop the contraction continuing until there is a singularity, a point of infinite density, at the center of the black hole.

THE UNIVERSE IN A HOLE

In the picture where our entire universe is in a black hole, the pre–Big Bang universe is vast, probably infinite, and stretching back indefinitely in time. Small fluctuations produce gravitational attraction between small amounts of matter, which are pulled together in clumps. Over the eons, these clumps become so dense that a black hole forms. But within the black hole the condensation goes on and on, until the matter is condensed to the order of magnitude of the strings that constitute it.

This intense compression results in a massive explosive expansion—the Big Bang—but all this takes place inside the confines of the massive black hole. In such a black hole universe we are permanently isolated from the "outside," but there is no reason why there can't be many other black hole universes out there, each permanently disconnected from the rest, unable to cross the black hole's perimeter and reach the true universe. In this case, the Big Bang becomes just one of many, probably an infinite set of universe-in-a-hole formations across the much larger true universe.

Things start to get even more interesting when M theory is brought into the picture because, unlike in string theory, all the extra dimensions need not be curled up in an incredibly small space, but could, as with the spatial dimensions we are familiar with in our universe, stretch off to infinity.

Is there any evidence of this? Well, possibly, although as with most cosmologically significant science, it is very indirect. Of the four forces we have already discussed, gravity is much, much weaker than the other three. It has been suggested that gravity can in some way ooze out of our three spatial dimensions into another part of M space. There's a kind of logic here. In general relativity, gravity is seen as a distortion of space, which implies an extra dimension in which that distortion can take place. It is possible that it is the leakage of gravity in this extra dimension that makes the force of gravity so relatively weak.

A CRASH OF BRANES

If our universe is a four-dimensional membrane (three of space plus time), floating in a cosmos of these extra dimensions, there is a fascinating possibility for a very different way for the universe to start. Instead of the explosion of a cosmic egg, it could be the result of a universal car crash. Imagine two membrane universes, floating in M space. They have no significant matter or energy; they are cold and dead. We don't know where they came from (that's a different issue) but this is before the Big Bang. The universe as we know it doesn't exist.

In this picture, gravity leaks out of the normal space of the universe within the brane, so the two membranes are attracted toward each other. Eventually they collide. A vast quantity of energy is produced when they hit each other. This becomes the heat and matter we see as the Big Bang from our position on one of the branes within the collision.

The two branes are now blasted away from each other. We are in one of those universe-on-a-branes. As our universe gradually cools and dies over the billions of years to come, it will end up much as it started. The force of gravity will take over again. Our universe will pile into its sister and everything will start up once more. It's a cyclical universe like the Big Bang/Big Crunch combination, but much more exotic as the cycle takes place in another dimension and brings in a second membrane universe as part of the process.

A neat feature of this model of the universe is that it does away with the complexities of inflation. The universe already was large, flat, and uniform, so there is no difficulty in explaining how it got into that state. And the dark energy driving the acceleration of the universe's expansion can be explained as the effect of the other membrane on our own. A rather frightening consequence of this description of the universe is that there's nothing to stop our membrane encountering another one that isn't its sister, resulting in a destructive crash long before we would otherwise expect the universe to end.

If this model were correct and we were heading for a collision, the outcome would be apocalyptic on an inconceivable scale. Not just the destruction of the Earth, or even of our galaxy, but the whole panoply of time and space as we know it would end, rendered into an inconceivably hot fireball. However, even if this is the case, we shouldn't give up just yet; unless we hit a rogue brane, the final collision should be billions of years ahead.

The physicists behind the bouncing branes theory, Neil Turok of Cambridge University and Paul Steinhardt of Princeton, believe that the timescale between collisions of the two branes is of the order of a trillion years. If this is true, with our current universe only around 14 billion years old, we shouldn't be looking out for another collision happening any time soon.

STARTING FROM SCRATCH

Turok and Steinhardt were inspired to come up with this theory because of the patched-up nature of the inflationary Big Bang theory. As astrophysicist John Bhacall put it when it was first announced that the results coming from the WMAP satellite were consistent with the inflationary Big Bang theory, "WMAP has confirmed with exquisite precision the crazy and unlikely scenario that astronomers and physicists cooked up based upon incomplete evidence. . . . Incredibly, everyone got it essentially right."

When M theory was developed, it was first thought that it might provide the explanation that theorists had been struggling for to explain just why inflation happened when and how it did. It seemed entirely possible that inflation could be driven by some sort of interaction of different branes, or with energy coming from the extra dimensions required by M theory. Unfortunately, nothing has yet emerged. It's possible something will, but this reflects the problem with M theory and string theory that has caused some physicists such as Peter Woit to refer to it as "not even wrong." There is potentially an infinite set of arrangements possible and there seems no way of showing how a particular arrangement is the right one.

One possible reaction to this situation is to move away from M theory, discarding it in favor of something else. But an alternative, if you believe as many physicists do that M theory still holds our best chance of having a single uniting theory that explains all the forces and particles we observe, would be to keep M theory and discard inflation. This was the option that Turok and Steinhardt decided to explore.

PARALLEL UNIVERSES

They were inspired by a lecture given by string theorist Burt Ovrut, who described a fascinating possibility. He imagined two branes, each of which could be a universe like ours in their own right, each occupying string theory's nine spatial dimensions, separated on the tenth dimension required for the united M theory but only separated by a tiny, tiny gap, as little as 10^{-32} meters (1 divided by a hundred billion, billion, billion, billion, billion). These branes are at the limit on the tenth dimension; in this picture, it goes no further.

In most respects, Ovrut suggested, there would be no communication between the brane that formed our universe and its parallel sister, but gravity could cross the gap. This might, he suggested, be an explanation for the effect that is described as dark matter, where parts of the universe are heavier than they should be given everything that's present, but we can't detect anything else. As we have seen, it is generally assumed that dark matter is composed of particles that have mass, but don't interact with electromagnetism, so we can't see them, but wouldn't it be neat if instead it were in a parallel universe, its influence felt in our own through its gravitational pull.

Initially this may seem a very arbitrary construction, as if a scientist had suddenly decided to say, "Let's assume the universe is a ham sandwich," but the parallel universe branes with the tiny gap between made a lot of practical sense. The particles we observe can be modeled as bubbles that sit in the gap on one of the branes. And the math of this parallel brane setup worked surprisingly well for much of the existing theory.

As Turok and Steinhardt listened to Ovrut, they were struck by one further possibility that he hadn't mentioned. What if the two branes collided? Couldn't this produce effects similar to those we associate with the Big Bang, but without the "beginning of everything" problems that dog that theory? Perhaps here was a mechanism to explain not only where our universe came from, but what it was like before the catastrophic fireball that brought our current collection of matter and energy into being.

Turok and Steinhardt believed that brane collisions were unavoidable in brane theory. Although it's not particularly obvious why they are unavoidable, it does seem reasonable that they could happen. Arguably, if the branes are really just 10^{-32} meters apart, and constantly interacting gravitationally, the more pertinent question might be why they don't happen much more frequently than the once-a-trillion-years event that Turok and Steinhardt have suggested. If there were a collision between the branes, the theory suggested it would fill them with an almost (but not quite) uniform distribution of both matter and energy, exactly what the real observations of our early universe suggest.

The essential unevenness required to establish the seeds of the galaxies we see today would also come naturally out of the nature of these branes. They would not have to be perfectly flat. If you imagine them instead to be flexible enough to have little wrinkles forming depressions and hills, then the collision would not happen at exactly the same moment all the way through the branes. Different points would be fractionally ahead of others, and that would be enough to generate the uneven kickstart we need our universe to have.

EXPANDING THE EXTRA DIMENSION

Although the realization struck the duo in one lecture, it then took many months of work to tackle the highly complex mathematics (some of it still not fully formulated) that would enable them to see what their model would produce in terms of the behavior of the universe with which we are familiar. In doing so, they moved away from the initial model with the extremely close branes, and went instead for two branes that were very far apart in another dimension. The tiny attraction between them would very slowly accelerate them toward each other.

They would start off in a low state of energy (as with an expanding universe that has reached the end of its life, flat and lifeless) but over the vast time they were heading toward each other, as the attraction between them grew and they got closer together, they would gain more and more kinetic energy, until at the collision there was enough to establish all the matter and energy we believe was in the post–Big Bang universe.

Turok and Steinhardt were delighted to discover that the ripples they expected would be generated by such a movement through a field of rapidly increasing strength were exactly those required to produce the sort of distribution of matter and energy in the resultant universe that we now see.

By this stage, they were ready to expose their idea to the world, and needed a name for it. Unfortunately, rather than go for something easily understood (and easy to read) such as "Big Bang" they went with the older scientific tradition of applying Greek tags and called this the ekpyrotic universe, meaning that it began out of fire, surely not ideal, both in its unfortunate look in English, and in the meaning, which hardly gives the suggestion of colliding branes. Sadly, not everyone in science can have the way with words of Gilbert Lewis who dreamed up "photon" or Murray Gell-Mann who came up with "quark." (Or even Fred Hoyle with "Big Bang.")

When they first presented the idea, Turok and Steinhardt had a problem pointed out to them. Their model assumed that the branes were flat and uniform when they started out long before the collision. How did they get in that state? They had initialization problems, just as those who supported the Big Bang and inflation did. Yet their worries didn't last long. They realized that this "problem" was in fact a way to get around the issue that had caused Hoyle, Gold, and Bondi to reject the Big Bang in the first place. There was no need for a beginning.

BACK TO CYCLES

Instead, the universe could be cyclic. That flat uniform state could be the end process of the stretching our universe is undergoing at the moment. Over a vast timescale, the branes could collide, stimulating a huge burst of energy, bounce apart, dispersing that energy as they stretched until once more they were flat and uniform, when they would be attracted back together, eventually collide, and the whole thing would go on once again.

With this model there was no need for a troublesome beginning (with the usual questions of from what? and caused how?). Instead the universe goes through an eternal cosmic cycle of attraction, bounce (with a spread-out bang), springing apart, and expansion until attraction takes over once more.

The dark energy that caused the acceleration in the expansion of the universe was no longer some mysterious unknown quantity, but the force at the heart of the brane–brane interaction. Although some of this energy appears to be lost to matter and light energy in the postcollision universe, the extra gravitational attraction from the new mass involved is just enough to keep the cycle continuing. Normal energy sources run out but we are not aware of any upper limit of gravity. It just keeps on going.

In fact the colliding brane model is almost unique in *Groundhog*

Day universes in being able to keep going indefinitely. Usually the second law of thermodynamics (see page 273) ensures that with time a universe that is contracting, going through a Big Bang, expanding to a limit, contracting again, and so on will eventually run down. It's thanks to some very special properties of branes that this isn't the case with this particular model.

One of the significant differences between Turok and Steinhardt's bouncing brane model and the old idea of a cyclic universe is hugely significant, but easily overlooked. In a way, it makes this model as much a relative of Gold, Hoyle, and Bondi's Steady State as it does of the Big Bang, because there is repeated creation in a potential infinite universe that is ever expanding; it just takes place over a much longer, punctuated timescale than the Steady State model suggested. Unlike the earlier cyclic models, it isn't our universe that is going through a cycle of expansion and contraction.

In Turok and Steinhardt's model the three physical dimensions we are familiar with only ever expand. It's the dimension that separates the two branes that expands and contracts. Not only does this do away with the problems with infinities arising from the infinitely small, infinitely energetic beginnings of the Big Bang, it means the universe is quite different in this model. At the end of our current cycle, as it heads toward brane collision, our universe will get no smaller.

At the collision, the two branes will once more generate lots of energy that will manifest itself as matter and light in the new universe, but the starting dimensions of that universe will be much bigger than this time around, and it will grow even bigger still, on and on, without any reason to end. In this picture, the universe we are aware of is only a tiny portion of the whole universe that is expanding; imagine that taking place on an enormous, perhaps infinite sheet, the brane. We only ever see a small local segment.

For those, like me, who once liked the Steady State theory and have never been happy with the Big Bang, this is a very appealing alternative. The only problem is that in doing away with the problems of the Big Bang, it does mean taking on the equally worrying M theory, which despite being, with its younger sister string theory, a very widely held scientific theory, is one that continues to have real problems in terms of being in any way established in experiment and measurable output.

This is not to say that the theory is a matter of good ideas and hand-waving. The mathematics behind the operation of branes is solid enough, and as yet it all works perfectly to make the bouncing brane concept of the universe a very realistic alternative to Big Bang plus inflation, but it has proved impossible to date to come up with a way of putting string theory or M theory to the test when setting it up against the real world. It fits with what we observe, but what it can't do is make a prediction we can then check against reality because there are usually too many possible outcomes.

EXPLAINING THE INEXPLICABLE

However, there is one potential advantage to Turok and Steinhardt's theory, and that is the immense timescale that may be involved. Although there is no prediction of how many times the cycle has already occurred, there is nothing to stop it having happened many, many times in the past. Now, as it happens, there is a measurement in cosmology that is drastically smaller than it should be. This is the cosmological constant lambda, the component that Einstein referred to as his greatest mistake, and that provides a measure for the influence of dark energy. If you put this down to the quantum energy of the vacuum, the usual explanation of why it exists, it is vastly smaller than it should be, by a factor of 10^{120} (1 with 120 zeroes after it).

Big Bang theorists have not been able to come up with an explanation of why this value is so small. It is thought that the value should fall over time but only very gradually. That would require the universe to be a great deal older than the 14 billion years we believe it to be. But in the bouncing branes model, that age limit is not a problem. The universe could easily be old enough for the cosmological constant to have decayed by this much. Although it's no proof, it is very satisfying that the bouncing branes theory takes away the most bizarre problem with Big Bang, one that generates the biggest mismatch between theory and practice anywhere in all of science.

Most of the measurements coming from the WMAP satellite studying the cosmic microwave background radiation have been supportive to both the Big Bang and to bouncing branes. Both would expect very similar results. So as yet there is no way to distinguish between the two, but the best hope for the future comes from measurements of gravity waves. As we will see, these should provide a very clear distinction between the two. But before we go into that in detail, there is one alternative approach to bouncing branes that has to be considered.

THE MANIFOLD DELIGHTS OF THE SLINGSHOT

Remarkable though the idea of colliding branes may be, one cosmologist believes that M theory can do away with the whole concept of a beginning, explaining the universe as we see it without any need for a collision of branes at all. It's a complex theory that requires a little more of a dip into the way branes might work, but it is wonderfully elegant because it removes many of the problems we find hard to explain with the current theory.

This is the brainchild of Cristiano Germani of the International School of Advanced Studies at Trieste in Italy. Like most M theory descriptions of the universe, Germani's envisages that the extra spatial dimensions we don't see are wrapped up in the beautifully named construct (sounding like something out of *Star Trek*) called a Calabi-Yau manifold. These also look beautiful when drawn as a 3D projection.

The trouble with Calabi-Yau space is that it tends to be unstable, twisting and wobbling, with openings blossoming and sudden spikes emerging. Germani wondered what would happen if the brane that formed our universe went flushing down the throat of one of these openings in what has been described as a slingshot universe. If the brane simply slid down to the bottom of the throat it would be crushed in a form of Big Crunch, but Germani imagined the brane would be spinning, rather like a sheet of paper pulled down a whirlpool.

A spinning brane in such a Calabi-Yau throat would bounce before it reached the very tip of the throat, heading back out toward the opening. As it headed out it would expand, giving the kind of expanding universe we see today. In this picture, the brane was not created at the end of the throat when it bounced; it had always been present. That means there was plenty of time prior to starting to expand when the different parts of the universe could even out so there is no need for inflation to explain why far-spread parts of the universe are so similar.

As yet, there is no measurement available that can distinguish between Germani's beautifully simple slingshot concept and a Big Bang plus inflation. Both agree with the observed results, except for the lack of gravitational waves (see page 155) which so far give more weight to the slingshot idea.

As with many other theories of cosmological uncertainties, there is hope that the next generation of cosmic microwave background telescopes such as the European Planck satellite will make fine enough measurements to distinguish between theories. In its favor, though, the slingshot model does do away with the singularity of the Big Bang, which stretches all theory past the limit, and also does away with inflation, as yet to have any sensible explanation for its beginning and end.

Germani's model is still very basic and doesn't yet explain all the details of forces and particles that have been possible with other M theory models that go into much more detail, however, it is a fascinating possibility that our universe has effectively been white-water rafting down a multidimensional waterspout.

QUANTIZING GRAVITY

If the universe really does go through such a repeated cycle of expansion and contraction, it is just possible that the uneven distributions in the early post–Big Bang universe that resulted in the formation of galaxies were leftovers from the universe before. One competitor to string and M theory in the attempt to pull together the otherwise incompatible relativity and quantum theory, called loop quantum gravity, also provides a model for how such a universe would work that has some similarities with the bouncing branes model, but that is dramatically different in approach.

Loop quantum gravity is an entirely separate theory from string and M theory that aims to provide a way to combine general relativity with space that is broken up into quantum units. On the whole it works well mathematically, but as with string theory there is a degree of arbitrariness in the way the math is applied to the real world. Even so, some physicists believe that loop quantum gravity has a better chance of accurately pulling together gravity and the other forces than any other existent theory.

In loop quantum gravity's picture of the universe, the fabric of spacetime is a tangle of local connections down at the quantum level, rather like zooming in on a piece of material with a very powerful microscope. (In fact, in this theory, when we talk of the "fabric" of spacetime, we are almost being literal.) When we take our normal, macro view of the world we don't see this frazzle of interconnection, just a smooth fabric.

Supporters of loop quantum gravity argue that, as a result of all these springy little connections, it predicts that a contracting universe will have a similar effect to crushing together a series of very elastic connections: when the pressure is removed, it will all spring back out again, a bounce that could provide a form of Big Bang without the universe ever reaching a troublesome singularity.

Early attempts to model a loop quantum gravity universe going through a bounce after collapsing suggested that nothing would remain from the previous universe. However, Martin Bojowald of Pennsylvania State University believes that spacetime itself became repulsive as it crumpled up. In this version of a loop quantum gravity Big Bang, there would admittedly be what Bojowald has called "cosmic forgetfulness" where the properties the universe had before the Big Bang are mostly wiped out and new ones are imposed. But Bojowald has suggested that the most detailed astronomical measurements could give us some insights into the nature of the universe pre-Bang.

This doesn't mean that we can see much of what came before.

Bojowald has said, "Some properties of the universe before the Big Bang may have only such a weak influence on current observations that they are practically undetermined." Bojowald doesn't envisage the scenario where the previous universe is quite like our own because so much is lost. "It's as if the universe forgot some of its properties and acquired new properties independent of what it had before," he has said.

REFLECTING THE PAST

However, when Parampreet Singh of the Perimeter Institute for Theoretical Physics in Waterloo, Canada, and Alejandro Corichi of the National Autonomous University of Mexico, Morelia, tried to model such a universe using the math of loop quantum gravity, they found that the process of going through a Big Bang bounce should not have a significant effect on the key parameters of the universe, those Goldilocks values that make it just right for life.

If Singh and Corichi are right, the universe before a loop quantum gravity bounce would seem surprisingly familiar. Our predecessor universe would have the same natural laws. Energy and matter would behave as they do in our universe. Time would also act much the same. "Since the pre-bounce universe is contracting," Singh has said, "it will look as if we were looking at ours backward in time." Because of the way very small differences result in huge changes in complex systems emerging from simpler ones, we would not expect the old universe to be identical down to the copies of you, occupying the same city on the same world, but on the large scale of galactic clusters there might be similar structures. Singh emphasizes, however, that galaxies could have formed in a different way in a predecessor universe, resulting in differences even at that level.

Unlike some possibilities for "what came before the Big Bang," Singh and Corichi's picture is very much "more of the same." Because they believe some degree of structure could survive the Big Crumple preceding the bounce, they think that we might be able to see a very limited picture of the structure of the pre–Big Bang universe in the cosmic microwave background radiation, that the strange patterning we see in the WMAP satellite might tell us almost as much about what came before as the structure of galaxies and galactic clusters that seem to have emerged from it.

Once you get down much below the level of galactic clusters, the opportunity for chaotic modification becomes too large. Very small differences that would inevitably occur in the variations in the early universe would get magnified over time and space, producing a vast butterfly effect, where a microscopically small difference would result in huge shifts in the final outcome.

However, at the time of writing, this idea isn't universally supported by those in the loop quantum gravity field (themselves in a relative minority among those trying to pull together a picture of how everything works). Not only are the models of the universe that Singh and Corichi used very simple, they aren't the only interpretation of the math, and others think there would be no continuity of constitution from the pre–Big Bang world. It could be that in the earlier universe those quantum tangles were magnified up to an environment where spacetime itself was similarly messy, a strange world indeed.

WATCHING GRAVITY WAVE

The jury is out. However, there is some hope of making observations that will allow a decision to be made between inflationary theories and ideas such as the Big Crumple or bouncing branes. As we saw on page 156, at the moment there is a surprising lack of evidence of gravity waves in the cosmic microwave background radiation. Models of the universe that involve a bounce rather than a start-from-scratch bang, whether they are based on loop quantum gravity or colliding branes predict much (much) smaller gravity waves being produced than would come from Big Bang plus inflation.

If we can clearly establish that the existing gravitational waves are much lower in intensity than those predicted by the Big Bang, it doesn't give us a way to select among bouncing branes, a Calabi-Yau slingshot, and a quantum gravitational Crumple but it does mean we can forget the Big Bang and start working on methods to select among the others (or we have to move to something completely different again). Unfortunately, this is not easily done.

Even at the level predicted by the Big Bang, these gravitational waves are incredibly faint. Nor can they yet be directly observed. Our best hope thus far is that they should result in slight variations in the intensity of the image produced by the WMAP satellite but there is the difficulty of distinguishing such variation from the simple quantum fluctuations in intensity that are thought to have produced the galaxies. If the distinction can be achieved, it will be by a statistical analysis, a bit like separating random noise (the basic energy distribution) from a signal (the gravitational wave) in broadcasting.

Another test can be applied that relies on polarization. This is a property of photons that usually has a random set of values, but that can take on a specific direction when light is absorbed and re-emitted by matter in the process known as scattering. (This is why Polaroid sunglasses, which use material that filters out light polarized in one direction, work. They cut out reflected glare, which tends to be polarized.)

The influence of gravitational waves on polarization is quite different from that of the overall energy distribution, so it should be possible to tell from polarization in the WMAP scans whether the variation is coming from differing energy density or gravitational waves. To date, the polarization shows no trace of a gravitational component, suggesting a very small intensity of gravitational waves and reinforcing the doubts about the Big Bang and inflation model.

However, once again we are dependent here on a very indirect measure. Although we do have a gravity wave detector in the form of LIGO (see page 172), this can only cope with waves that are much stronger than those that remain from the beginnings of the universe. Even the proposed space version of LIGO, the LISA satellite-based system with its measurement arms millions of kilometers long, would not be powerful enough to detect such waves directly, by a factor of about a thousand.

The information is certainly there in the WMAP data but so could be many other influences that would have a similar effect on the data. What we are seeing could be gravitational waves, enabling us to separate the Big Bang from its competitors, but it possibly could be something very different. In any case, to date, there is no supportive data for Big Bang, although the results aren't sensitive enough to rule out every possible model that is based around a Big Bang/inflation combination.

In going to more detail, there remain significant problems in detaching the effect of gravitational waves from the many influences on the cosmic background radiation that could modify it. After all, this stuff has been floating around for getting on 14 billion years, passing through a universe with plenty of matter to interact with it. However, astrophysicists are confident that with a good enough probe, they can bring the level of uncertainty down low enough that they can definitely say whether the gravitational waves that should have been produced by a Big Bang are present.

When the next generation Planck telescope produces much more detailed cosmic microwave background plots, should there still be no obvious distortions caused by gravity waves, then it will rule out the Big Bang plus inflation pictures more firmly and add support to the bouncing or scrunching views for what led into our particular universe. It's not certain whether Planck will get down to quite enough detail to be sure, however. Cosmologists are pushing for a dedicated satellite to concentrate on the polarization issue, making measurements that will be more cut and dried, but at a time of concern about science spending, we may have to hope that Planck can deliver the goods.

Whatever the outcome, it may be that the picture being used by all the protagonists in this chapter, including supporters of the conventional Big Bang, is too small scale, because it only features the one universe (or two, one on each of the colliding branes). Some cosmologists believe that our universe is just one of many.

10.

LIVING IN A BUBBLE

The most valuable lesson to be learned from the history of
scientific progress is how misleading and strangling [analogies
drawn from the history of science] have been, and how success
has come to those who ignored them.
—THOMAS GOLD (1920–2004),
Cosmology (in *Vistas in Astronomy*, ed. Arthur Beer)

If we go with the existence of the Big Bang in the most widely accepted version of the theory, we also need inflation (see page 136),
the phenomenal sudden expansion of space from practically no
size at all to something more in the scale we are used to thinking
of for a universe.

If this did occur (and we have to remember that inflation is an
add-on to fix a problem with the original theory) that inflation
might not be a one-off. Some scientists have suggested that it
could happen again, or even that it could still be happening on a
regular basis. If this is the case, the first universe (not necessarily
ours) could have produced other "baby universes," budding off
the original like the kind of plants that reproduce by producing
Mini-Me versions of themselves.

LIVING IN A BUBBLE

The advantage of this localized view of inflation is that it is possible to be more flexible about its nature. The most commonly suggested reason given for inflation to stop is that it was an unstable process that naturally decays. But being a quantum process, we would expect variation in this decay and, in principle, inflation could go on forever. Different bubbles within the whole universe (usually called a multiverse to avoid confusion with the single bubble that is our universe in this picture) would stop inflating at different times. A new bubble could provide the sort of environment that we consider to be our universe, whereas other parts become newly stretched areas where once more the quantum slate is wiped clean. This has the advantage of being able to answer, "Why did inflation stop when it did?" with, "It didn't, really."

Leaving aside what's going on in all those bubbles (we come back to that in a moment), the most dramatic thing about this bit-by-bit inflation is that unlike the traditional, one-off Big Bang universe there need never have been a beginning to the multiverse. It could have been going on forever, bubbling out new universes, forming unstable bubbles at the edge of the universe where inflation has stopped and bubbling out into yet more new universes. Our own universe could in principle be the first but there could equally have been an infinite set of predecessors, with ours just one in a vast branching tree.

If this is the case, it's quite reasonable that we can't see any evidence for the other bubbles. We can only see as far into space as time allows us (around 13.7 billion light-years if current dating is correct) but there is no reason why the entire bubble that we occupy should not be much bigger than this. Current estimates allow for our universe to be anything from around 50 billion light-years up to well over 100 billion light-years and in principle our bubble could be vastly greater than this but we could never reach another bubble without finding some way to travel faster than light.

VARYING PARAMETERS

For some cosmologists, the thought of these universes existing is not just a reasonable one, but the most likely. And if those extra universes exist, then we can't assume that they are all like our own. The nature of our universe is established by a series of natural parameters: the speed of light, for example, or the size and nature of atoms. There is no particular reason why these parameters should be the same in other bud universes (although, of course, they might be) and it could be that many universes exist where matter as we know it never came into being, whereas many others could have a much more exotic form of matter and perhaps life.

In isolation, the idea of multiple universes—forming a multi-verse—although entertaining, has little impact on us. These hypothetical universes may make for good fun in speculating what a universe would be like with, for example, a slower speed of light or a higher gravitational constant, but as long as we can't communicate with them in any way, they don't have any real influence on us. In fact, about the only scientific value of a multiverse with many different physical parameters is that it gives one possible explanation for the "Goldilocks" nature of our own universe.

As mentioned on page 17, scientists have long been surprised by the fine-tuning of all these constants required to allow a universe that could permit the production of life as we know it to exist; all these factors, like Goldilocks' favorite chair in the three bears' cottage: not too big, not too small, but just right. Even a small variation in many of them would make it impossible for life as we know it to have formed.

In Chapter 2 we saw how this principle can be used to suggest that the universe has a designer who made it specifically so we can exist. But there are two other possibilities.

The first is simple chance. Ours just happened to be the kind of universe that emerged, and inevitably it's the kind that supports our life, or we wouldn't be here to write (and read) books about it. We wouldn't exist. But we do, and so that's how it turned out. Period. No implications at all. The weak anthropic principle (see page 19) again. Some people find this highly unsatisfactory. It's so incredibly unlikely. Think of all those other universes that might have been, they would say; the chances that this one was what happened are tiny. They would argue that it just can't have happened by chance.

COPING WITH CHANCE

However, this is a misunderstanding of chance. As human beings we are terrible at handling probability and chance. We just aren't programmed for it. Take a simple example to see just how bad we are at this. To celebrate the New Year, a lottery has decided to run a double event. Two draws on the same day. Two chances to become a millionaire. As usual, stacks of tickets have been sold. Millions of people are holding on to the hope of having their lives transformed. Of course, they know it's unlikely. They probably won't win. But there's a tiny chance that they could soon be living the celebrity lifestyle, and that's what makes spending the money worthwhile.

The first draw goes much as any other week. Out come the numbers, one after another, a random string of possibilities and hopes. This particular lottery uses six numbers from a possible forty-nine. There's a bonus ball too, but that's for losers. Let's watch those six big numbers: 24–39–6–41–17–29. It's over for all but the lucky few. There's nothing exceptional. Although the announcer tries to inject wild excitement into his voice, it has been done a thousand times before. It's not news.

Then the second draw begins. Let's pick up the commentary.

> They're using the machine they call Delilah for the draw to-night, started by a former member of Blondie. Here comes the first number and . . . what are the chances of that? The first number is one. Okay, next choice. Well, would you believe it? Two. That's incredible. And the third number's coming through now. Hey, is this a joke? The third number is three . . .

And so it goes on until the draw is complete: 1–2–3–4–5–6. All neatly in sequence. There is uproar. The payout is suspended while the draw machine is overhauled and checked. A trickle of demands for a refund soon becomes a torrent. Questions are asked, all the way up to the government. Yet no one can find anything wrong. How could this be? How could such an incredible result happen?

That simple lottery draw exposes a strange disturbing reality. Our world has many random elements in it, where probability is the only guide. Probability is an essential contributor to what's happening in our world. Yet human beings are incompetent when it comes to handling the outcome of chance. We just don't get probability. It seems unnatural, and fools us all the time. There really was no reason to be surprised by the lottery draw. The outcome 1–2–3–4–5–6 is just as likely to come up as 24–39–6–41–17–29. It has exactly the same chance. Yet to our probability-blind brains there is a huge difference between the two results.

It might seem strange that we can't cope with probability if it's so important, but evolution often produces a compromise, where one capability is sacrificed to make another strong. The ability that makes it impossible for us to handle probability well is pattern recognition. We depend on patterns. They provide our interface with the real world. So strong is our need for patterns that we frequently make them up where they don't exist. Where there is no pattern, where probability rules a sea of randomness, we are lost.

This is why the lottery result takes us by surprise and why so many people bother to buy lottery tickets in the first place. When we see a draw such as 24–39–6–41–17–29 our probability blindness conceals just how unlikely it is that a particular combination of numbers is going to be drawn. It is only when a pattern is imposed and we see 1–2–3–4–5–6 that we realize just how improbable the whole thing is. As we have seen above, patterns are essential to help us cope with the world but our enthusiasm for patterns blinds us to the impact of probability.

So our inability to grasp probability fools us into thinking that a lottery draw of 1–2–3–4–5–6 is special. It's the same with the universe. All of those constants and positioning of the Earth and so on have to have some value. Okay, the chance that a particular set of values comes up is very small, but no more than any other combination. The particular combination means that all sorts of other potential life forms didn't come into existence and get surprised by how things are. It could be pure chance it was "our universe" that came along.

EACH ONE UNIQUE

The final possible reason for the existence of a Goldilocks universe is one of particular interest when trying to answer what came before the Big Bang. What if there are billions of different bubble universes, each with its own values of all the physical constants? Some with no Earth at all, some with an uninhabitable Earth, and just a few, of which ours is one, with the Goldilocks settings of the constants. That way, we wouldn't be anything special. Our universe would just be one of many, many others all existing in parallel. Some cosmologists believe this is how the universe is.

Others go even further and equate the concept of a multiverse where universes "bud" off other universes with the processes involved in biology. If the right circumstances apply (and we come back in a moment to what these are), you could imagine that the explanation for our "Goldilocks" universe is that universes have evolved to be that way. Just imagine that a universe budded off a range of universes, some more stable, others less stable than the original one. It would be the more stable universes that stayed around long enough to bud other universes, so gradually, a higher and higher percentage of universes would have a stable form.

Is this idea, dreamed up by the physicist Lee Smolin, who has such problems with string theory, a realistic one? He suggests that universes might be born from within black holes, but unlike the earlier universe-from-black-holes theory, the new universe doesn't appear within the black hole, or even within our universe, but in a new, otherwise inaccessible chunk of spacetime. It is possible that such a mechanism would allow the new universe to carry through some but not all properties of its parent universe (where the black hole is) with it.

For evolution by natural selection to work there need to be a number of mechanisms in place. You need reproduction, you need some properties to be passed on to the offspring, with randomly introduced variation, and you need the capability to reproduce to be limited by survival factors. How do our breeding universes match up against these mechanisms?

Clearly if they can bud, universes have the means to reproduce, and the capacity to reproduce will be limited by how stable the universe is. But will properties be passed on from one universe to another? After all, this is essential for evolution. They may if Smolin's idea of the origin in black holes in the parent universe is true. And would there be random variation? Quantum theory pretty well guarantees that.

However, it should be pointed out that there are plenty of stable configurations of a universe that are incapable of supporting life as we know it, or unlikely to produce the right circumstances for life. By most ways of thinking, all an evolutionary process would produce is stable universes. But there is one, pretty far-out, way that evolution could encourage the ability to support reasoning life.

That strange idea relates to the fact that quantum events are influenced by an observer. It is arguable that a universe with life forms might develop differently than a universe without life forms because there was a different class of observer within the universe, so quantum processes would happen in different ways. Similarly, a universe with intelligent life forms might develop differently from one without intelligent life forms. If this is the case, having intelligent life could be a survival trait for universes, so once arrived at by random chance, it could be more likely to continue in future universes.

PHANTOM MULTIVERSE

I ought to make it clear that not every cosmologist and astrophysicist is a supporter of the multiverse idea. If you read some books on the subject you might think this is an idea that has as much support as the Big Bang and inflation (within our own part of the multiverse if necessary), but that's not the case. Many scientists are unhappy with it. Paul Steinhardt, one of the pair behind the bouncing branes theory (see page 208) has said, "This is a dangerous idea that I am simply unwilling to contemplate."

He calls it dangerous because he believes it causes confusion and is "pure fantasy." However, if we allow the possibility, there are two problems that must be tackled. The first is that there is already a different possible type of multiverse arising from quantum physics (although the quantum version is often called the multiple worlds theory to try to limit confusion). And the second is that there's the danger of this being an unprovable theory, and many would suggest that such a theory isn't science at all.

THE QUANTUM MULTIVERSE

Quantum physics, the science of the very small, is full of mysteries that, unlike those in cosmology, are eminently capable of being studied by experiment. At the heart of quantum theory is the idea that a quantum particle is fuzzy. Instead of having a clear position, for example, there is only a range of probabilities for where it is. So if you fire a photon of light—the definitive quantum particle—at a pair of slits it behaves in a way that seems impossible in the real world. It goes through both slits at once, and will only settle on a single slit if you actively measure which slit it went through.

Many quantum theorists just accept this as the nature of reality at the level of these tiny particles. In the above example, the photon is literally in two places at the same time unless you make a measurement. But others find this so counterintuitive that they believe every time a quantum particle faces a "decision" such as which slit to go through, instead of having both values, each with a certain probability, there are two separate universes: one where it goes through the left-hand slit, the other where it goes through the right.

This idea comes in two flavors. In one version, an infinite set of universes already exists: in some of these the photon goes one way, and in some the photon goes the other (in many more, the photon never existed, or missed the slits entirely). The second version of the quantum multiple worlds theory suggests that each time such a quantum event occurs our universe splits into two, one for each of the possible outcomes. That way there is no need for the photon to spookily decide which way it went if a measurement is made; it's just that the measurement takes place in that particular universe.

Both flavors of this theory tell us we live in a multiverse or "multiple worlds": the first flavor with a preexisting infinite set of variations on our universe, and the second which is constantly splitting. Given there are probably well over a googol (1 with 100 zeroes after it) quantum particles in the universe, many going through such events on a very regular basis, that's a whole lot of universes to split off. This quantum multiverse is quite different from the cosmological idea of a multiverse with separate Big Bangs in different bubbles. The cosmological and quantum multiverses could exist in isolation of the other, or we could have both simultaneously, a multiverse of multiverses.

FIRST CATCH A BLACK HOLE

The more serious problem with the cosmological multiverse theory is that it could be impossible to prove or disprove. As with many beliefs, this ends up as something other than science. It is, in effect, theology, and so has no real place in cosmological science. To take a ridiculous example, according to Douglas Adams' superbly funny *The Hitchhiker's Guide to the Galaxy*, somewhere out there in space, the Jatravartid People of Viltvodle Six firmly believe that the entire universe was sneezed out of the nose of a being called the Great Green Arkleseizure. Now there is no way science can prove that this has more validity than one of the real human religious ideas of the origin of the world. That doesn't mean that Adams' joke is on a par with real religions, but it's not something on which science can comment.

If the idea of multiverses is just as untestable as the Great Green Arkleseizure because there is no way to find even indirect evidence, then there is no basis for calling it a scientific theory. However, it clings in there because it's possible there may be some evidence we can eventually discover, and because there is even the possibility that we could communicate with another universe in the multiverse. This is a nontrivial task. If it were to appear in a recipe book, the first instruction would be "Catch yourself a black hole."

We have already met black holes as a physical concept, but let's think for a moment about the practicalities of dealing with a black hole, if we ever managed to get our hands on one. Black holes are astronomical bodies from a science fiction nightmare. If you flew toward a black hole in a spaceship, the force of gravity would get bigger and bigger. Movies like Disney's *The Black Hole* that portray ships flying into a black hole miss a disastrous consequence that was highlighted in an Arthur C. Clarke short story called "Neutron Tide" (the story featured the less dense neutron star, rather than a black hole where the effect would be even greater).

Clarke pointed out that as an object (in the case of this story, a monkey wrench) got closer to the star, the difference in the force of gravity between one end of the wrench and the other would be immense. The end closer to the star would be pulled in so much harder than the other end that the wrench would be stretched out like a piece of spaghetti. As the story concludes, all that would be left is a star-mangled spanner. The same would happen to any ship or person.

Worse still, at the event horizon itself (the point of no return beyond which nothing escapes) if the equations are correct, the pull of gravity becomes infinite, and this has a particularly strange outcome, thanks to general relativity. Einstein's *special* relativity says that the closer in speed an object gets to the speed of light, the slower its time appears to be to an outside observer. *General* relativity goes a step further and says the stronger the gravitational pull a body experiences, the slower time appears to be from the outside. If we watched an object traveling into a black hole it would get slower and slower before stopping forever at the event horizon. It should take an infinite amount of time (as far as the external observer is concerned) for the object to get through.

When these conclusions were first drawn, with no suggestion that this was more than the workings of the numbers, it was simply assumed that they showed that such black holes could never exist. Relativity generates quite a lot of possibilities that are generally considered impossible because of the implications. If an object (or even a signal) could travel faster than light, for example, it would move backward in time. This was considered just another example of the strangeness of the potential implications of relativity that would never crop up in the real universe.

This was certainly Einstein's view on the subject. As with the probabilistic nature of quantum theory, a branch of physics that emerged from his work, he was not happy with the infinite gravitational force at the event horizon of a black hole, and even when a different type of mathematics was applied that did away with the infinite gravitation he could not accept that a black hole was a real body. Instead he took it as a mathematical curiosity that fell out of general relativity.

WORMHOLES IN SPACE

Ironically, it would be Einstein himself who with associate Nathan Rosen in 1935 realized that in principle black holes could act as tunnels to another location. The term they used was a bridge (still referred to as an Einstein-Rosen bridge, although better known by the colloquial term "wormhole"). Einstein was trying to use the theory behind black holes to explain fundamental particles, and imagined that to avoid the unfortunate physical consequences of a true black hole existing, they might come in pairs that counterbalanced their properties.

This was what would later be called a black hole/white hole combination (where a white hole is a kind of anti–black hole), forming a wormhole—an instantaneous link—from one point in space to another. In principle, and we come back to this, such links could even join one universe in a multiverse to another. Even more than black holes it should be emphasized that we have no evidence that wormholes in space truly exist; they are just a construct that is allowed by the mathematics and may exist, or may be possible to force into existence if they don't occur naturally.

The fact that a black hole could, in principle, be a portal to another location seemed, however, to be one of those interesting but unusable results that often seem to arise out of modern physics. Although black stars could in principle act as a bridge from one location to another, it seemed a shortcut that it was impossible to take in the real world.

First there was the time aspect. From the viewpoint of an observer outside the black star's gravitational field, any traveler trying to cross the event horizon should get slower and slower, but never actually make it through. It has been pointed out that this relativistic effect doesn't apply to the ship itself. Time on the ship will seem to progress normally so the observers on board would be aware, if they could survive, of the ship plunging into the black star gateway.

However, what is less often pointed out is that this relativity paradox seems to imply that the ship can never get out of the wormhole and return to its starting point. If it did, when it joined the original observer, it would be in two places at once. From the newly returned ship's viewpoint, the original version of it would still be ever more slowly approaching the event horizon; those on the ship would be able to see themselves. Of course, arguably, they are seeing only a delayed image so this isn't an insurmountable obstacle.

What's less easy to brush aside is the impossibility of surviving at all. There is no point entering a black hole gateway if in the process you are stretched into spaghetti, and even without the gravitational field going infinite, that "star-mangled spanner" effect would still be plenty big enough for the differences between the two ends of the ship to pull it apart. This is a barrier to practical application that would not be even theoretically surmountable until around the time that physicist John Wheeler came up with the graphic term "black hole" to describe what had previously been called dark or frozen stars.

BLACK HOLE ROULETTE

Wheeler first used the term "black hole" in 1967, and it was just four years before that mathematician Roy Kerr not only came up with a way of passing through a black hole's wormhole without being shredded, but even showed that his idea was more likely than the traditional picture of a black hole as universal destroyer. It's all a matter of rotation. In space, on the whole, things rotate.

Think about the solar system for a moment. We know the Earth spins around its axis as it travels around the Sun. The Moon spins too, at just the right speed to keep the same face always toward us. (This isn't coincidence: the Earth and the Moon influence each other, and over time have been pulled into synchronism.) It's the same for the Sun too. That spins around its axis, and so does every other star for which such observations have been made. In some cases we can even measure the speed of rotation of distant stars, as in the pulsars discovered by Jocelyn Bell (see page 118).

So imagine a black hole is forming. A huge dying star is no longer reacting strongly enough to fight against the immense gravitational pull of all the matter in it. It gets smaller and smaller until it flashes through the limiting radius where light speed is overcome by the gravitational pull; it has become a black hole. Now unless it was very unusual, this star would initially have been spinning. As it gets smaller and smaller its spin will get quicker and quicker.

This is basic physics, the conservation of angular momentum. It's the same as someone spinning around on a playground merry-go-round who suddenly goes from hanging out as far from the center of the ride as he can, to pulling up close to the center. Or a pirouetting ice skater who suddenly pulls in her arms. Suddenly they begin to spin much faster. Similarly, as the matter in the star collapses inward, the speed of the spin will increase until the surface is traveling incredibly quickly.

Kerr discovered that in such circumstances, rather than stretching a ship entering it until it is ripped apart, a rapidly spinning black hole would enable it to stay in one piece as it passed through the hypothetical wormhole and appeared out of the white hole at the other side. Remember again, however, that the existence of black holes is not proven and wormholes doubly so.

There are plenty of "ifs" and "buts" for this picture. The exit of the wormhole could be nearby, or could be in a different universe within the multiverse. And although the shredding effect could be avoided, there would still be plenty of effects from the acceleration applied to any item that went through a black hole gateway that may well be totally impossible to encounter and stay alive. This is not something we are likely to try out any time soon, and much though science fiction fans yearn for it, wormholes are not likely ever to provide us with gateways to the stars.

EVIDENCE FOR BLACK HOLES

Although I have stressed several times that the existence of black holes isn't proven (we haven't yet got good enough imaging to show the ultimate blackness of the event horizon of a black hole or the radiation that should be given off as matter accelerates into it) almost all astronomers and cosmologists are convinced they do exist and have some convincing indirect evidence for both black holes of around the mass of an ordinary star, and massive black holes with the equivalent mass of millions of stars, which tend to be found at the center of some kinds of galaxy, including our own Milky Way.

What have been captured, particularly since we've had space telescopes and very large radio telescope arrays, are images of what is thought to be the swirl of space junk on its more distant route into a black hole. Bearing in mind the black holes are expected to be spinning, it's a bit like a whirlpool in water, a spread-out disk of matter that is whirling around on its way into the center. From the speed of rotation of these disks, it is thought that the unseen object at the center has such a huge mass that it would have to be a black hole.

Oddly, we can see the effects of such "ordinary" black holes easier than the immense black hole that is thought to be at the center of our galaxy. There is so much dust between us and the center that we can't see anything, otherwise, if it were truly there, we should see an ever-present glow as large amounts of matter were accelerated into oblivion.

We do know that many galaxies are rotating fast enough to suggest a very heavy concentrated mass in the center, and although it isn't obvious why this should be due to a supermassive black hole rather than yet another variant on the dark matter theory, this is the most common explanation. Similarly, some galaxies have radiation pouring from their centers, which is thought to be the radiation produced as matter is sucked into the hole. Assuming there is such a black hole at the center of the Milky Way it is thought to be as heavy as 3.7 million Suns.

If the mass that influences galaxies' rotation were due to dark matter rather than a black hole, it would be necessary to explain why radiation comes bursting out of these galactic centers. In 2006, Anatoly Svidzinsky of the Texas A&M University in College Station, Texas, came up with a reason. Some of the models of dark matter suggest that giant bubbles would emerge from a cloud of dark matter on a regular basis, spewing out radiation just as the central black holes appear to do. Even so, black holes probably do exist, and if they do, they just may be able to provide a gateway through to another part of the multiverse.

If such a multiverse of bubble universes exists, the chances are that we aren't part of the very first universe, and as such we have a satisfying answer to what happened before the Big Bang. The event we think began everything was only the beginning of our universe, and in the bubble multiverse there were many earlier Big Bangs.

DETECTING THE MULTIVERSE

As we have seen, there is no evidence yet of there being other universes out there, but there are some possibilities for the future. Some versions of multiverse theory suggest that there is at least one other bubble universe very, very close to our own, perhaps only as little as a millimeter away. If that's the case it is possible that some of the effects, particularly gravity, "leak through." As is suggested in the bouncing branes model, this could be responsible for the production of dark energy, or dark matter, but those who speculate that this kind of universe exists are waiting eagerly for the Large Hadron Collider (LHC) at CERN to be fired up.

CERN (Conseil Européen pour la Recherche Nucléaire) is a vast international research establishment working on high-energy particles, nominally in Geneva but in fact straggling over (or, rather, under) the border between Switzerland and France. Best known now for the massive spin-off success of its electronic communication vehicle the World Wide Web, and its role in Dan Brown's thriller *Angels & Demons*, CERN is a place where the basic components of the universe are battered together with immense energy in an attempt to analyze their makeup and understand their characteristics.

CERN has been operating for many years now, but its greatest device, the LHC, is just starting to be fired up. LHC is an eighty-four-kilometer-long underground racetrack, a huge metal tube along which particles are accelerated faster and faster in a nightmare carousel, spurred on by vast magnets that take the power of a good-sized city. (When I give talks on science and global warming, I have been asked several times how it's possible to justify CERN's power usage and hence CO_2 emissions. I tend to reply that it's easier to justify than the power used in watching reality TV.)

When the particles in the LHC racetrack are traveling at close to the speed of light they will smash into each other, causing an incredibly high energy reaction, similar to the temperatures involved at the Big Bang (although on a much smaller scale). This is science at its least subtle, the sort of approach small children tend to take. Hit something as hard as you can and see what happens. Ask scientists working at CERN what the most exciting outcome will be, and they are likely to say the potential to discover the Higgs boson. This is a hypothetical particle that gives the other particles their mass. Discovering it would give a great deal of support to the current "standard model" of how particles are made up.

Others would say that the most exciting outcome is also the most terrifying. A few people have suggested that the LHC could bring the end of the world, or even the end of the universe. This could be the outcome of re-creating the Big Bang or, through the production of a bizarre particle called a strangelet that some think could start a runaway chain reaction, converting all of the matter in the planet to a different form. A failed attempt was made to take out an injunction against CERN to prevent them from destroying the universe.

CERN plays down the risk, emphasizing that even the kind of energy produced in the LHC is small compared to natural events, and that strangelets (unstable particles made of strange quarks, a different kind of quark to those that make up our protons and neutrons) are only hypothetical and wouldn't transform the world into strange matter if they were produced. In practice, no one knows exactly what will happen, but the feeling is that the end of the world is relatively unlikely to emerge from switching on the LHC.

However, cosmologists believe that there may be something even more exciting than Higgs bosons produced by the Large Hadron Collider: tiny, particle-sized black holes. According to the best current theory, such nano–black holes could not be produced with the sort of energy levels the LHC can generate, but could only come into being if a parallel universe were providing extra gravitational input. If they do appear, they will generate a lot of interest in the existence of other universes in near-reach of our own.

IS ANYBODY OUT THERE?

Could we go further still and not only detect other universes but communicate with them, or even travel to them? Physicist and broadcaster Michio Kaku believes so. His reasoning goes something like this. After billions of years, we will reach a stage where our universe is no longer a suitable place for life. Many projections of the future are of a gradual rundown of the universe until it is a cold place, lacking in energy, where even some incredibly evolved life form based on pure energy would be increasingly slow and tend toward death.

Bearing in mind this end would be billions of years in the future, Kaku speculates that our distant descendants (or another, later intelligent form of life) may, indeed, find a way to travel from one universe to another, surely the ultimate proof of that other universe's existence. To be able to make it through, this civilization is likely to have to find naturally occurring wormholes in space (see page 238), or to be able to create them.

As we have seen though, wormholes may not exist or be feasible to create, and might not be usable even if they can be made real. And there should be another word of caution here. Although Kaku is a respected scientist, responsible for parts of string theory, he is also heavily influenced by science fiction, and *Star Trek* in particular. When I spoke on Kaku's radio show a while back, he was vaguely irritated that I wouldn't accept that time travel was unlikely in the foreseeable future. He wanted this science fiction standard to be true a lot more than the scientific evidence suggested was likely. The same is sadly true of his speculation of interuniversal travel.

The fact is that we know that reality contains a number of insuperable barriers. There are, for example, some mathematical problems that have been proved can never be solved. It could also be that there are some dreams of science, such as faster-than-light travel, time travel, and bridging universes that will never be possible, no matter how advanced the civilization. But Kaku has hope on his side, and it would be churlish to deprive him of his dreams.

If we do live in a bubble, then, it might be that our picture of a single universe beginning with a Big Bang is wrong, but we could never have a way of proving that to be so. However, a bubble universe isn't the only way we could be deceived about what's around us, and therefore about what came before the Big Bang. It could be that everything we experience is an illusion.

11.

WELCOME TO THE MATRIX

No one who has experienced the intense involvement of computer modeling would deny that the temptation exists to use any data input that will enable one to continue playing what is perhaps the ultimate game of solitaire.

—JAMES LOVELOCK (1919–),
Gaia: A New Look at Life on Earth

How do you know everyone else and all the stuff around you exists? How do you know that the whole universe is not in your head? It's a thought that occurs to most teenagers, fitting well with their naturally self-centered viewpoint at that stage of their development. It has also occurred to philosophers in the past, who call this view "solipsism," from the Latin for "alone" and "self."

In one sense solipsism is very appealing. We are always encouraged to select the simplest explanation that will explain all the observed facts, ever since the medieval friar William of Ockham came up with this principle, which is known as "Ockham's razor." In one sense, solipsism is a very simple explanation for the universe as it only requires one mind. However, it has the downside of presenting us with a very complex universe for such a simple source: it's hard to see why a universe that is the construct of a single human mind would require the complexity we observe. Arguably, you would expect a solipsistic universe to be much more like the places we inhabit in dreams, very limited in scope.

IT FROM BIT

In this extreme solipsistic view, the whole universe is inside your head. In such a view, the Big Bang was just a stage of your mental story. Although this seems very "soft science," quantum theory does offer one possibility for the universe to be literally constructed around you, and this is only one of the ways to fulfill a *Matrix*-style model of the universe, where the whole creation is a construct (electronic or otherwise) and pre–Big Bang existence can be just what you want it to be. Such an unreal universe may seem unlikely, but John Wheeler, who was one of the most respected physicists working on cosmology, believed it was possible.

Wheeler was born in Jacksonville, Florida, in 1911 and took his first degree at John Hopkins University in Baltimore. He spent most of his senior career at Princeton, except for a ten-year stint as director of the Center for Theoretical Physics at the University of Texas at Austin. Wheeler worked on quantum theory (the great Richard Feynman was one of his students) but is probably best known now for his contributions to the more dramatic aspects of astrophysics. He also came up with some of the better-known names in the field, coining both "black hole" and "wormhole."

Wheeler's dramatic theory about the nature of the cosmos says that the universe itself is nothing more or less than information, just like the bits in a computer. His "it from bit" theory has a self-kickstarting universe where the act of observing brings information into life. In a sense, Wheeler's universe only came into being as a result of being observed and it's even possible that it could describe a "young earth" creationist view, where the world as we know it came into being 100,000 years ago when human beings' brains became sufficiently capable of observation and analysis that a complete universe was required.

Leaving aside this bizarre possibility, Wheeler's theory is still intriguing, suggesting as it does that everything we experience is just a stream of data. This is different from a true *Matrix*-style world, where the whole universe we experience is a simulation running on a real physical computer. In the "it from bit" theory, the universe is the information; there is no computer.

REGISTERING ISN'T BEING

Some would argue that there is a danger of ridiculous oversimplification when using this theory. Take a statement about the "it from bit" theory, given by quantum computer scientist and "it from bit" supporter Seth Lloyd: "The universe is made from bits. Every molecule, atom and elementary particle registers bits of information." There is an unwarranted logical jump in that statement. Just because something registers bits of information doesn't mean that it is made from those same bits.

Take an everyday object, a TV remote, say. I can say that object registers one bit of information depending on whether it's held button-side up or button-side down. I could send you a message that way. If you come into the lounge and I've left the TV remote button-side up, then it could mean "I've made dinner"; if it were button-side down it could mean "Please go out and buy dinner." Because it can indicate only two things (in this particular mode) it registers a single bit of information, which a computer would think of as 0 or 1, but I have given a rather more complex interpretation linked to a meal.

However, that bit of information is not related to how the object is constructed. The remote is not made from that bit; it just registers it. Similarly, saying that properties of fundamental particles can be allocated values that correspond to bits of information is not the same as saying that they *are* those bits of information. You might as well say that I don't need the latest MP3 player, as I have the complete technical specifications the manufacturers use to build it, and I should be happy to own and listen to that information, rather than the player itself.

However, leaving aside that unfortunate oversimplification, it is still possible to conceive of the entire universe being, Wheeler-style, a collection of information that is constantly being modified by the interaction of these bits. And as, at this level, we are dealing with quantum physics, these are quantum bits. One of the interesting results that falls out of this is that the behavior of the universe is only predictable in a probabilistic way.

REDISCOVERING FREE WILL

Compare a universe of quantum bits with the universe as Isaac Newton saw it. He also had a kind of "it from bit" vision of the universe, but for Newton the universe was a mechanical computer, like a massive version of the Difference Engine dreamed up and partly constructed by Victorian visionary Charles Babbage. In Newton's mechanical universe, if you knew every bit of data about the current state of the universe, you could, in principle, predict how it would behave forever.

Newton himself did not express this idea, but others did.

Pierre-Simon, Marquis de Laplace, was a French mathematician and scientist working at the end of the eighteenth and start of the nineteenth centuries. He was, without doubt, a major contributor in both science and the math that would prove essential as physics became more mathematical, but in the materialist spirit of the French enlightenment, he was convinced that it would require a complete knowledge of the universe at any point in time to be able to predict it absolutely in the future, including human behavior; for Laplace, free will was a meaningless concept.

I say that we could predict how the universe would behave "in principle" in Newton's universe because even in this "it from mechanical bit" universe, there would be so many bits that it would take an impractically immense calculation. But even so, in principle it is doable. However, because at the level of quantum particles we can't specify the exact state of all the components of the universe, but only apply probabilities, we can't ever predict the behavior of a single particle in detail, let alone the whole universe. In fact, to have a computer able to model the entire universe, that computer would have to be another universe in its own right and even then, it would soon diverge, as probabilistic measurements came up with different results.

YOU CAN'T DUPLICATE PROBABILITY

To see what I mean about diverging, imagine a very simple probabilistic universe that consists of three coins, all with the heads facing up. My "computer simulation" of that universe is another set of three coins, also all faceup. It's a "computer" that is just as big as the original "universe," and contains the same number of bits of information. Now let's set both the "real" universe and my computer simulation running. We toss all the coins several times, alternating between a toss with the real universe coins and the simulation.

At the end of the process, the "real universe" might be Heads–Tails–Tails, whereas the simulation is Tails–Heads–Tails. Even at that simple level, the simulation hasn't stayed with the original. Note that I'm not saying that the universe is entirely random, but it has probabilistic aspects, like the coin tosses. This random aspect of the quantum bits that make up a Wheeler-style universe may seem frustrating, but at least it can give us something closer to free will than the programmed inevitability of a Newtonian mechanical universe, where everything has to follow, cycle by cycle, from the initial settings.

This doesn't mean that it's impossible to make any predictions in a quantum universe. If I take a ball and throw it at the wall, I can say with some confidence that it will bounce off the wall and fall to the floor. Quantum theory tells me that there's a very small probability that it will pass straight through the wall, and an even smaller probability that it will disappear and emerge on the other side of the world. These outcomes are so unlikely that I can ignore them for all practical purposes. But taken at the level of individual quantum particles, I do have to take account of potentially strange behavior, including tunneling straight through apparently solid barriers.

Of itself, the "it from bit" theory doesn't really change anything about the way we view the universe. All it is saying is that the universe can be regarded as a set of quantum bits, and the operation of the universe can be thought of as a vast computation using those bits, as an alternative to the traditional physical theories of how the universe works. Really, it's not an "it from bit" theory at all, but a "bit from it" theory. It doesn't of itself say that physical things are made from bits, but rather that bits can be made from physical things.

Such a picture really doesn't give us any useful route into the question of what came before the Big Bang, because practically speaking the "it from bit" universe is identical in nature to the conventional physical picture, and all the options discussed in the previous chapters apply equally well. However, once you have accepted that every detail of the universe can be considered as a quantum bit, you have opened up a route into a potential situation where "it" is very much from "bit."

PROGRAMMED LIFE

Wheeler's idea does not require a computer, because the universe *is* the computer, but we also need to consider a world that has much more in common with *The Matrix*. I don't mean the literal and ridiculous storyline in this otherwise impressive (and fun) movie. Using human beings as energy sources is a ludicrously inefficient way of generating energy, particularly with all the overhead necessary to keep the artificial world they inhabit in place. However, it is a perfectly acceptable supposition that the world as we know it is a vast computer program, run on machines built by an intelligence we know nothing about, in a universe that could be like any of the ones we speculate about, or one that is totally strange and alien.

It has been argued that this couldn't be done with sufficient accuracy because of chaos and complexity. When different bodies interact, they do so with a complexity that increases incredibly quickly, making exact calculation of the outcomes virtually impossible. Even a simple astronomical problem of deciding how three objects in space will interact is too messy to simulate with absolute precision. Imagine how much more complicated it is with the billions upon billions of items in the universe.

To make matters worse, chaos theory tells us that very small variations in the input of some kinds of systems (the weather system of the Earth, for instance) can result in hugely different outcomes. So different that they become impossible to predict in a relatively small time. That's why long-range weather forecasts will always be much worse than predicting tomorrow's weather.

What's more, there are some problems, quite simple problems such as choosing the best route from A to B on a complex road network, that are impossible to solve exactly with any conventional computer. This has been mathematically proven. We can approximate a solution, which is how computer-based routing software works, but cannot solve these problems exactly.

Some argue that this makes a *Matrix*-style universe impossible. They say it would be impossible to handle the complexity of the system. We can't even model one human being, let alone the billions on the planet.

However, these issues aren't necessarily the showstoppers they appear to be. Firstly it might not be necessary to model with such accuracy. After all we can't tell if the universe is behaving incorrectly if we can't handle the computation. An approximation that gives the impression of being right would be enough. And we are limited by our current inability to use quantum computers. With a quantum computer, the sort of calculations necessary to run a universe may be possible.

ENTER THE QUBIT

Quantum computers are devices that use quantum particles—photons, for example—instead of bits on a silicon chip to do their computation. In principle, a quantum bit (called a qubit, pronounced "cue-bit," for short) can process infinitely long numbers. Where an ordinary bit can only deal with 0 or 1, the settings of the qubit are its quantum properties, such as its spin or polarization. The polarization of a photon can be in any direction at all around the axis of its movement. To represent it exactly as a number would take an infinitely long decimal. It's pretty obvious such quantum computers should be able to do amazing things. But there is a huge problem attached to getting quantum computers to work.

Although the information is there in a qubit, it's very difficult to get anything into it or out of it. If you measure the polarization, you only get one of two values, parallel to the direction measured or at ninety degrees to it. It's as if you get a 0 or 1 out of what could be 0.739012891 . . . The probability with which you get parallel or perpendicular is what we think of as the direction of the polarization; this has an infinitely long decimal value. But we can't measure that; all we can get out is parallel or perpendicular.

It's a bit like watching a game of pool on a black-and-white TV. All the information is in the pool hall in the real world, but you can't get to it through the screen; you can only see the indistinguishable gray balls. And quantum computers are all extremely tricky to handle, because any interaction with the bits spoils the values. However, if a full-scale quantum computer can be built, we already know some of the things it will be able to do, because, perhaps surprisingly, some of the programs to run on such computers have already been written. Just a couple of examples of these show just how much more a quantum computer can do than the everyday variety.

The first case of quantum computing brilliance involves a quantum version of the search for a needle in a haystack (the original paper describing this method was titled "Quantum Mechanics

Helps in Searching for a Needle in a Haystack"). The algorithm (a set of mathematical rules that in this case can be used only on a quantum computer) was devised by Lov Grover at Bell Labs and provides a way to speed up unstructured searches immensely.

Anyone who has used a phone book knows that it's easy to look up a number if you know someone's name. That's a structured search, because the phone book contains lists of names in alphabetical order. But try finding to whom a particular telephone number belongs. Then the task becomes much more difficult. Imagine you had a phone book with 1,000,000 listings in it. You might have to look at 999,999 entries before getting to the number you wanted. On average you would have to check 500,000 before finding the right number.

With Lov Grover's quantum algorithm, you only need to look at the square root of the number of entries (in this case 1,000) to be sure of finding what you want. This amazing speedup of searching becomes more and more essential as we deal with the complex messes of information with which our increasingly connected world presents us. There comes a point when any conventional computer will take too long to do a search through unstructured data. But a quantum computer could do it in the square root of the expected time.

Variants on the same method allow complex routing problems beyond the capabilities of a program such as Streets and Trips to be handled easily. The way route planning software gets to the "best" route is always an approximation. It's impossible for a conventional computer to come to an absolute solution of a complex problem of this sort in the lifetime of the universe, but quantum computers would find it trivial.

Another application of quantum computing that already has the algorithm waiting for the hardware to run it on to be built has computer security experts trembling. This is the ability to break down a huge number into the two prime numbers that have been multiplied together to generate it. With big enough numbers, this problem is beyond the capability of any conventional computer we can envisage. But a quantum algorithm already exists to crack the problem, if only there were a quantum computer on which to run it.

Why should we care? Because the encryption used on every computer—for example, when you see a little padlock to tell you that you're safe when you enter your credit card number into a Web browser—is based on a technique that relies on the difficulty of breaking down huge numbers into a pair of primes. If you can work out the primes involved, you can crack the code. Most current computer security would fall apart.

This isn't the only application of this ability to deal with primes, but it emphasizes the frightening potential of the quantum computer. At a stroke, if a quantum computer could be built, it would be able to solve a problem that the whole IT industry assumes is insoluble.

Because quantum computers can, in effect, handle infinitely long values, they could in principle model a real universe to a level of detail that would mean we would not be capable of detecting the difference between the model and reality.

BREAKING DOWN THE UNIVERSE

Quantum theory also comes into another way of looking at the feasibility of the universe being a computer program. Where the sort of traditional classical physics we learn in high school is continuous, meaning that just representing one straight line requires an infinite set of points, quantum physics is, effectively, digital. It suggests that there is a minimum physical size, below which it is impossible to go, the so-called Planck length of around 10^{-35} meters ($\frac{1}{10^{35}}$ meters, where 10^{35} is 1 with 35 zeroes after it).

If that's the case, we can imagine the whole universe as a grid of these Planck pixels, which makes it possible to limit the state of the universe to a finite number of bits, in principle each with a calculable value. Even if we don't resort to this kind of limitation, there's a lot our hypothetical universe programmers could do to simplify the problem. The extreme picture of this is what you might call the "solipsistic dream" version of the universe.

LIVING IN A DREAM

Imagine you were dreaming right now. Within the limits of the single, admittedly very sophisticated, wetware computer that is your brain, you have temporarily constructed a universe. Admittedly it's not a very good universe by most scientific standards. Things keep shifting. The scope to explore is limited. Even so it's an existence. So the very simplest form of a computed universe is one that's inside your head every night.

Because we "know" we exist, but not for certain that anyone else does, it is possible to imagine a computer-based world built around the sort of solipsistic picture with which we started the chapter. Outside the little bubble of what you know directly, everything can be suitably fuzzy, filled in only where necessary. What do you know about healthcare in Byzantium? Not a lot? In that case, there *is* no information about healthcare in Byzantium. It doesn't exist, and it won't unless you go and look it up or are "accidentally" exposed to it, at which point some computing power needs to be assigned to making it real.

Does it sound complicated, simulating all those billions of lives and feelings and thoughts out there across the world? Don't worry. It's only the people you have contact with that need any degree of detail. Everything else can be broadbrush, little more than a TV news or newspaper report. It's a bit like the movie *The Truman Show*, in which Jim Carrey unwittingly lived in an artificial world for a soap opera, except here there is no need to hurriedly construct physical objects and bring in people; it can all be done automatically by the universal computer environment.

A computer-based universe with just you in it in any detail is the minimal view, but there is no reason that, given a sophisticated enough quantum computer, we can't extend this picture to include as much of the universe as you would like. Here we truly have "it from bit." There is no independent existence for you, or other people, or the can of soda that sits on your desk. Each aspect of the universe from the simplest air molecule to the atomic structure of your brain that somehow generates your consciousness, to a quasar in the far reaches of the cosmos is just an array of bits.

THE ILLUSION OF MOVEMENT

In this world, physical movement is illusion. Nothing moves as it seems, just as when the bits in a conventional computer produce a moving image on the screen there aren't objects really moving around in space. This ties in rather neatly with a series of paradoxes dreamed up by the Greek philosopher Zeno over 2,500 years ago. We've already seen Zeno's paradox of Achilles and the tortoise (see page 33). The reason he dreamed up this story was because he belonged to the school of Parmenides, who believed that all motion was an illusion.

The best example to illustrate his attitude to motion is the paradox of the arrows. He describes an arrow flying through space. After a certain amount of time has passed, that arrow will have moved to a new position. But now let's imagine it at a particular instant in time. The arrow must be somewhere. You can imagine it hanging in space as it would in a single frame from a movie. That's where the arrow is at, say, exactly ten minutes past two.

This is where the visual imagery of film comes in handy. There is now a video technique (first widely used, by coincidence, in *The Matrix*) that seems to make time stop. An object freezes in space as the camera pans around it, showing it from different directions. (In fact what is happening is that a series of cameras at different angles capture the moment, and their images are linked together by a computer to produce the illusion that the camera is panning.) Imagine that we do this for real. We stop time at that one instant and view the arrow.

Now let's do the same for another arrow that isn't moving at all. We won't worry too much about how this second arrow is suspended in space. If it's really a problem for you, we could work the paradox with two trucks, one moving and one stationary, but Zeno used an arrow, so I'd like to stick with that. The question Zeno asks is, How do we tell the difference? How does the arrow tell the difference? How does the first arrow know that it must change positions in the next moment, whereas the other, seemingly identical in our snapshot, stays still?

In our universe in a quantum computer, the only difference between the two arrows is a parameter, at least at the basic level. Arrows, we know, are made up of atoms, each with its own collection of unique quantum parameters. In principle, the two arrows would be different at the detailed atomic level as well as in their relative velocity.

However, our computed world could cheat here again, and only fill in that level of detail when required. Interestingly, quantum theory says that the very act of observing quantum particles changes them. Given we accept that being observed has an influence on real objects, you could say that such a "lazy" quantum computer–hosted universe, which only filled in the details when required, was what we might expect from our understanding of the quantum world.

REWRITING HISTORY

If the universe is truly in an environment like this, then anything in the far past, such as the Big Bang, is tantamount to fiction. It need never have existed in the "real" timeline, but could simply have been constructed whenever the simulation began running. The universe could have started yesterday, last month, last year, or 14 billion years ago as appears to be the case. Note that in such a situation, we are dealing with something very different to the sort of fake we experience in the normal physical world.

When a few years ago diaries were discovered that appeared to have been written by Adolf Hitler, there was great excitement. Subsequently they proved to be forgeries: good imitations, but clearly distinguishable from the real thing. If our universe were created yesterday in a quantum universe simulator, then there would be absolutely no difference between this and it actually having existed for 14 billion years. The evidence in both cases is just an identical collection of quantum bits. It isn't a fake, it's the real thing, but one that needn't have actually "lived" through the time period required for its history.

If we are in such a simulator, pretty well anything could have happened before the Big Bang (especially bearing in mind that our universe might not have existed yesterday). It could be that the "real" external universe is similar to our own, or completely different. The simulator could have been built by intelligent beings (the next best thing to gods as far as we are concerned, but beings that could be just as natural as us) or it could have emerged naturally. The simulator could be run numerous times. There could even be multiple levels of simulation: the computer our universe runs on could be itself a simulation on another computer.

CHANCES ARE WE'RE SIMULATED

It seems highly unlikely that the simulated universe we know would spring into being fully formed, but if you combine the universe simulator with a multiverse, with each version of the multiverse different by just one quantum bit, eventually you will hit on the universe that is the way we experience it. In fact some have argued that if you accept the multiverse idea it is almost inevitable that there are computer-based universes out there. Somewhere among the universes there would, in all probability, be universes where civilizations had developed far enough to produce a *Matrix*-style universe. And if you believe this for our existence, then chances are that we do live in a computer simulation of a universe.

If universe simulators do exist in a multiverse, then any one planet could house many, many simulations. There could easily be many more simulated universes than there are "real" physical inhabited universes. So, on a purely probabilistic argument, if you accept the concept of a varied multiverse you also have to accept that it is more than likely that the universe we inhabit is part of a computer simulation. If you really think the multiverse contains many universes inhabited by intelligent beings, the chances are that you are just an artifact in a computer program.

Although a quantum universe simulator throws wide open what came before the Big Bang, it does also open up the possibility of communicating beyond the simulator. Just as a normal computer interfaces with the external world through screens, keyboard and mouse, webcam, microphone, and speakers, so our universe simulator could have the means to communicate with the external world. We may never find it, or it could be discovered tomorrow. For that matter, the beings running the universe could pop in and say hello. Whether we could distinguish them from gods is a different matter.

This is in many ways a highly unsatisfactory universe, just because it is so flexible and open, but then that's the nature of a universal computer.

Such visions of living within one individual's brain, or as part of a vast computer, can really stretch the mind and get us thinking differently. But serious physicists have come up with the even stranger picture of reality and its origins with which we finish.

12.

SNAPSHOT UNIVERSE

The idea of an indivisible, ultimate atom is inconceivable to the lay mind. If we can conceive of the idea of an atom at all, we can conceive of it as capable of being cut in half; indeed, we cannot conceive it at all unless we so conceive it. The only true atom, the only thing which we cannot subdivide and cut in half is the universe.

—SAMUEL BUTLER (1835–1902),
The Notebooks of Samuel Butler (ed. H. Festing Jones)

What do you think of when you hear the word "hologram"? The rainbow tints reflecting off a security marker on a credit card? A spookily three-dimensional green image that's like a window onto a real view? The 3D projection of Princess Leia by R2D2 near the start of the original *Star Wars* movie? They are all (with the possible exception of the last) about producing a three-dimensional view out of a flat two-dimensional surface.

Some physicists believe that the whole universe is holographic in form, that the three spatial dimensions we think we experience are actually all part of a very sophisticated projection: not the security marker of an intergalactic bank, or a strange 3D movie show of the gods, but an unexpected facet of nature that means the three dimensions of space we believe we experience don't actually exist. If this is true, it totally transforms the nature of the universe, Big Bang and all. Before seeing what is involved, though, we need to understand a bit better just what a hologram is.

THREE DIMENSIONS INTO TWO WILL GO

The hologram is a good example of an idea that came along before the technology that made it possible to construct. All the holograms we see today depend on using lasers to produce them, but the idea of the hologram was dreamed up by Hungarian-born British scientist Dennis Gabor nearly twenty years before the laser became a reality. Soon after the Second World War, Gabor was thinking about the way we see objects. It's something so commonplace that we take it for granted, but many of the best ideas in science and technology come from taking a closer look at something that's apparently everyday.

Imagine looking through a glass window at a coffee cup on a table. Stand over to the left and you see a certain view of the cup, perhaps the handle and the front side. Move around to the right and the view gradually changes, taking in different angles of the three-dimensional object. All the light required to make up these different views is falling on the window glass. So if there were some way to take a snapshot of all that light, of every ray (or rather, every sequence of photons) traveling from the cup to the glass, you should be able to re-create the view from the window, with an image that changes as your viewpoint does.

To cope with all the photons coming from different directions you would need to distinguish not just how bright a particular point is, as an ordinary photograph does, but also what the phase of the photons is, a property of the photon that changes with time and that corresponds to the position the wave is in if you take a wave view of light.

To do this, Gabor imagined using a second beam of light falling straight on the glass. The two beams, the one bounced off the coffee cup and the other directed onto the glass, would interfere with each other like beams of light passing through a pair of slits in the Young's slits experiment often done at school. Perhaps the simplest picture of interference is if you imagine dropping a pair of pebbles into a smooth still pond. Each pebble will set off a circular set of ripples, heading outward.

For a while those two sets of ripples will flow out into the water independently, neither having any effect on the other. But eventually the outer ripples will collide. When this happens, some will reinforce each other. If both ripples are heading up at the same time, the result will be a stronger ripple. However, if one ripple is heading up at the same time as the other is heading down, they will cancel each other out and the result will be an area of water that hardly moves at all.

The overall result of interference is a pattern with different intensities, spread across the pond. The same thing can happen when two light beams meet. Luckily, however, it's not so easy with light. Photons of light aren't particularly enthusiastic about interacting with each other. Most of the time they ignore each other. If this weren't the case we wouldn't be able to see, use a cellphone, or watch TV.

If you could see trails of light left behind by the photons that are crisscrossing the room in front of you right now, they would be heading in all directions, constantly passing through each other without noticeable effect. In that one room (or for that matter in your backyard if you're reading this outside) there will be visible light heading in all directions as it reflects from objects all around you. There will be radio, TV, and cell phone signals passing in all directions. There will be other short-range nonvisible light, from Bluetooth connections to wireless networks. An incredible mess if they could collide and interact, but they don't. On the whole they pass through each other.

However, if you get the two beams of light just right, then they will interfere, just as the ripples from those two stones in the pond. For this to happen the light beams have to have the same frequency (or same energy if you think of them as photons) and the phases of the two beams (see page 152) have to stay in the same relation. This is why when you do the interference experiment many people experience in high school, you don't just shine a pair of flashlights at a screen. You shine a single color light, preferably a laser, at a pair of very narrow, close-up slits. This makes it as close as possible to matching the requirements for interference.

Interference is why you will often get dead spots with cell phones or wireless networks, where reflections (at the same frequency and in the right phase) interfere with the original signal. A hologram is, in effect, an interference pattern between the light from the objects and the second "reference" beam that falls directly on the glass. The resultant pattern indicates just what phase each photon of light was at when it hit the glass.

BRINGING THE VISION TO LIFE

When he devised his three-dimensional pictures, Gabor's intention was to improve the electron microscope by enabling it to produce an image that could be seen from a range of directions.

Gabor's science was always driven by an immediate practical application. In his teens he had built a sophisticated laboratory at home with his brother George. There they went well beyond building a crystal radio set to constructing X-ray devices and experimenting with radioactivity.

With this practical bent, Gabor originally studied engineering rather than physics (reasoning that there were many more jobs available for an engineer), but he attended college in Berlin at a time when great names like Einstein and Planck were active, where his practical drive was tempered by an interest in the underlying science. In the case of his 3D microscope, however, the practical result escaped Gabor.

Even when taking the simpler approach of using light rather than electrons, there was a problem: Gabor couldn't make one of these pictures (they were soon called holograms from the Greek *holos* meaning whole and *grapho* to write), because they would work only if the light came from a special kind of source that didn't exist, a source where all the light was in phase. Once the laser, which is just such a source, was produced in 1960 the theory was all ready to be put into practice and it only took four years before Emmett Leith and Juris Upatnieks at the University of Michigan produced the first true hologram, a bizarre still-life of a model train and a pair of stuffed pigeons.

Early holograms such as Leith and Upatnieks' also had to be viewed using a laser, but more modern holographic techniques have produced holograms that can be viewed in ordinary white light and the sort of thin-film reflective holograms used to authenticate credit cards, bank notes, and DVDs. In these, light is sent through a plastic film from behind using a reflective metallic layer; it is the effect of the hologram on this light that produces the 3D image.

SEEING 3D

Readers of a certain age might feel that holograms are nothing new. They had 3D pictures on their View-Master way back in the 1950s and 1960s. There was a time when you couldn't visit a famous location without seeing those cardboard disks with the pairs of color transparencies around the rim.

View-Master was a compact version of the Victorian equivalent of TV, the top parlor entertainment of the times, the stereoscope. The stereoscope is a rare example of a technology that flourished over a hundred years ago yet still seems remarkable today, because it is no longer commonplace, and because its successors have never had the same commercial success. Instead of one photograph, a stereoscopic camera takes two simultaneously, with a separation that parallels the separation of the eyes, so each picture presents a subtly different view. We see a three-dimensional view of the world because the brain combines "flat" images from the two eyes, separated by around two and a half inches, and uses their slightly altered viewpoint to detect depth. The two simultaneously taken images in a stereograph are viewed through the stereoscope, which presents one picture to each eye, allowing the brain to combine them in the same way.

At one time it was thought that the optic nerves that link each eyeball to the brain were connected purely into opposing sides of the organ (so the right-hand side of the brain received signals from the left eyeball, and vice versa), but this would make it very difficult for the brain to make the high-speed comparisons necessary for stereoscopic binocular vision. It was discovered in the 1990s that in a fetus, although all the neurons growing from the eye begin by crossing toward the opposite side of the brain, some are repelled by a special protein and end up connecting to the same side of the brain as the eye.

The more growing neurons that bounce back from the midline of the brain, the better the binocular vision. Around 40 percent of our connections don't cross over, as opposed to 3 percent in mice, and none at all in many birds and fish that totally lack binocular vision. Because each side of the brain, handling its opposing eye, has information from the other eye to compare, between them they can build up a comprehensive picture that gives us a convincing illusion of seeing in three dimensions. Each half of the brain still handles information primarily from the opposite side of the field of view, but this information comes from both eyes.

The idea of presenting a different image to each eye to get a more realistic picture predates photography. Back in the late sixteenth century, Giovanni Battista della Porta and Jacopo Chimenti da Empoli produced side-by-side drawings, and in 1613 the French clergyman François d'Aquillon came up with the term stereoscopic (or to be more precise, *stéréoscopique*); but without an optical device to control the view of each eye it is very difficult to focus the two eyeballs separately, and the limits of accuracy of drawing by hand made the whole thing impractical until photography was invented. However, it was just a matter of years after the earliest true photographic image was made before a working stereoscope was built.

The device that was to become so popular in Victorian households was invented in 1838 by Charles Wheatstone, the British challenger to Morse in the attempt to build the first electric telegraph. Wheatstone gave it both of the names that would attach to it over the years—the stereoscope and the stereopticon—although he never went beyond a theoretical description of the process. It wasn't until 1849 that David Brewster made use of the fledgling photographic technology to build the first stereoscopic camera. It featured twin lenses, separated just like a pair of eyes, to capture a pair of images on the photographic plate behind it.

Just as a celebrity endorsement might raise awareness of an entertainment product today, it was Queen Victoria's enjoyment of the stereoscope at the Great Exhibition, held in the purpose-built Crystal Palace in London in 1851, that turned the stereoscope from a novelty to a huge commercial success. Within five years, over a million homes in the United Kingdom owned a stereoscope, and this enthusiasm for three-dimensional pictures was soon reflected across the Atlantic in the United States where the technology was promoted by Oliver Wendell Holmes.

Stereoscopes went on to have a long and checkered history. In later years they became widely employed in aerial photography, which since the mid-1940s has depended on the extra information provided in a stereoscopic image to make it easier to identify features on the ground. The stereoscope as entertainment saw its final outing to date in the View-Master system we've already met. The viewer changes the image on display by pressing a lever to rotate the disk. These toys were hugely popular in the second half of the twentieth century, and although less common now, are still available from the retail toy-making giant Mattel.

These days, we are more likely to come across 3D at the movies. Ever since TV proved a threat to movie theaters the movie business has been looking for ways to make the experience more dramatic. Although there has been a range of technologies employed, whether it's using two different colors for two images, or images that are polarized in different directions, they all rely on projecting a pair of images simultaneously, which are separated for the eyes by special glasses.

Such stereoscopic images are vastly different from a hologram, the difference between looking at a photograph and looking through a window. A hologram is a three-dimensional view. A perfect hologram would be indistinguishable from looking through a piece of glass at reality frozen in time because every photon is captured as it hits the viewing screen, and re-created when the hologram is projected. The View-Master and its movie equivalents fool the brain into thinking it is seeing in three dimensions by presenting subtly different views to the two eyes. But those views are flat. It's just a representation of three dimensions, whereas a hologram is a true three-dimensional view, captured and stored on a two-dimensional surface.

PROJECTING REALITY

In principle there is no reason why the universe cannot be like this too, appearing to have three spatial dimensions (or more), but in reality occupying less, all acted out in a holographic form on a lower-dimensional stage. It might seem that this idea falls down when faced with Ockham's razor, which you will remember suggests we should take the simplest view that explains the circumstances rather than add unnecessary complexity.

Tantalizingly, though, there are some aspects of the math behind the formation of the cosmos that work better in the reduced dimensional form, and there is some evidence from the theory of black holes that suggests this isn't such a hare-brained idea.

The starting point is the understanding that there is a correspondence between the universe and information, something we saw in the previous chapter in the "it from bit" theory. Here we consider the properties of each quantum particle to represent quantum bits of information in a vast universewide computation.

Now let's go back to our black hole. Imagine I take an object—

perhaps a monkey wrench, as in Arthur C. Clarke's story—and throw it into a black hole. Once the object reaches the event horizon it ceases to exist. The information in that wrench seems to have been lost, as is its capacity to hold information. Now, this is a problem, because there is a fundamental law of the universe that seems to make it a very uncomfortable possibility for the capacity to hold information to disappear. It's called the second law of thermodynamics, and some scientists believe it is the most fundamental of all physical principles.

INFORMATION CAPACITY CAN'T BE DESTROYED

The astrophysicist Arthur Eddington, whom we have met a number of times already, once said:

> If someone points out to you that your pet theory of the universe is in disagreement with Maxwell's equations [the equations that describe how electromagnetism works]—then so much the worse for Maxwell's equations. If it is found to be contradicted by observation—well these experimentalists do bungle things sometimes. But if your theory is found to be against the second law of thermodynamics I can give you no hope; there is nothing for it but to collapse in the deepest humiliation.

It's the second law of thermodynamics that makes perpetual motion machines impossible; one implication of it is that you can't get something out of nothing.

This is often phrased in terms of entropy, but that's quite a confusing term, so we show it instead in terms of order. (Entropy is a measure of the lack of order. The higher the entropy, the more the disorder there is.) Roughly, the second law of thermodynamics says that the level of order in a closed system will stay the same or decrease, but can't spontaneously increase. This is sometimes used (very badly) as an argument against the validity of evolution, or for the existence of God. It's worth taking a look at that argument, because it helps illustrate what the second law of thermodynamics is really about.

The anti-evolutionary argument goes like this. According to evolution, we start off with a random soup of chemical ingredients and over time this evolves first into single-celled animals, then more complex animals. Yet this is prohibited by the second law of thermodynamics, which says order can't emerge from chaos, so evolution must be wrong, and a creator must have done the arranging. Unfortunately for those who use it, this argument misses a fundamental part of the law: it talks about a closed system. That's one where things can't get in and out from the outside. The Earth isn't a closed system; we are constantly supplied with energy by the Sun, and that energy more than compensates for the increase of order involved.

The second law of thermodynamics doesn't prevent us from losing useful information. If I have a set of children's blocks with letters on them spelling out a word, I can decrease the order by scrambling up the blocks and lose the word. But what I don't lose here is the capacity for information. The blocks still have that. In fact a scrambled set of things has the capacity to hold more information than an ordered set. (Think of two slightly different sets of blocks, a "more ordered" one where all the blocks are the same color and a "less ordered" set where each block is a different color. I could use the colors to distinguish between the blocks and so could use the second set of blocks to store more information than I could the identical blocks.) What the second law of thermodynamics doesn't like is if the capacity to hold information at the fundamental level goes away, because that's like going from less ordered to more ordered.

So it isn't acceptable that matter's ability to store information is lost when it goes into a black hole. As the black hole eats up matter, the hole grows. And what theorists realized is that the only way to keep the second law of thermodynamics happy is if the event horizon, the theoretical sphere that describes the limit of the black hole, could hold all the information that went into the black hole. In this way, as the event horizon grows, its growing capacity for information balances out the capacity lost from the matter that is eaten up.

REALITY PROJECTED ON A SPHERE

Now there's just one more step to the logic required to leap from this theoretical black hole to the real world. Imagine we'd got a lump of matter, whether it's that monkey wrench or the whole Earth. It will contain a certain amount of information capacity in all the qubits of all the quantum particles that make it up. Now let's imagine that we turn that object into a black hole by intense compression. At the end of that process, we'll end up with a black hole that is smaller than the original matter. The information capacity of the black hole's event horizon must be as big as the original item. So the information capacity of all that matter can be described on a sphere that's smaller than it. That means it takes less area than the surface area of the object to contain the information capacity within it. The two-dimensional surface of a sphere is capable of holding all the information held in a volume bigger than the one it contains.

This means that there is no reason why our world, which apparently has three spatial dimensions, cannot actually be a two-dimensional collection of information we just happen to perceive as inhabiting three dimensions in space. In practice, it's a bit more complicated. The math doesn't work in a universe that is expanding into a limitless space, as ours appears to be. Yet there have been more complex variants on the world-on-a-surface idea that do seem to apply to our kind of universe.

Of itself, such a projected holographic universe doesn't tell us anything about what came before the Big Bang but it may make it possible in the future to come up with a different way of exploring back through the limits of the Big Bang. And it would be hugely important, as it would be saying that everything we think we understand about the universe is based on a false premise, meaning we would have to take a very different approach in exploring the nature of the universe.

Remarkable though this is, there is an alternative view of a holographic universe that is even more dramatic in its separation from reality as we generally see it. The holographic universe we have seen thus far still uses the conventional interpretation of physics, but projected onto a different number of dimensions. But one of the world's leading quantum physicists came up with a quantum model of the universe where nothing is as it seems, and the whole concept of dimensions is little more than an illusion.

THE QUANTUM MAVERICK

The man in question was David Bohm, an American scientist, born in Wilkes-Barre, Pennsylvania, in 1917. Bohm got his doctorate under bizarre circumstances at the University of California, Berkeley, during the Second World War. Because he had left-leaning political interests he was not allowed to join the Manhattan

Project to work on the atomic bomb with many of his colleagues. However, his doctoral dissertation covered a subject of significant use to the Manhattan Project, so it was immediately classified and he wasn't allowed to present it or to receive his doctorate. Luckily, Robert Oppenheimer, who headed up the Manhattan Project, had been Bohm's supervisor and was able to get Berkeley to accept that the dissertation was a success without it ever being officially read.

After the war, Bohm moved to Princeton, but by the end of the decade his political affiliations were getting him in trouble again, coming under the spotlight of McCarthy's House Un-American Activities Committee. After nearly two years of stressful consideration he was acquitted, but by now he had lost his job. Bohm left the United States and never returned to a university there, spending time in Brazil and Israel before settling in the United Kingdom where he would work for the rest of his life.

Bohm's field of expertise was quantum physics, the physics of tiny particles such as photons of light and electrons. During the 1930s, quantum theory had been developed to become a very powerful tool for exploring and predicting the results of the way these particles behaved. One aspect of it, for example, quantum electrodynamics (see page 200) has proved to be the most accurate theory ever in terms of its prediction of what would be observed in experiments. Yet at its heart, quantum theory has some philosophical problems that caused Bohm real concern.

These were the same philosophical problems that had caused Albert Einstein, whose early work had been absolutely essential to the development of quantum physics, to doubt the theory's correctness. This led to a series of arguments between Einstein and Danish physicist Neils Bohr, who was the lifetime champion of the interpretation of quantum theory that is still most widely used.

GOD DOESN'T PLAY DICE

It was this difficulty with quantum theory that spurred Einstein to his most famous remark:

> Quantum mechanics is certainly imposing. But an inner voice tells me that it is not yet the real thing. The theory says a lot, but does not really bring us any closer to the secret of the "old one." I, at any rate, am convinced that He is not playing at dice.

This has been often condensed to "God doesn't play dice." The problem he had was that quantum theory was fundamentally based on the idea that everything was driven by probability. We can't even predict as simple a thing as where an electron that is traveling through space will be a little later in time. We can only assign probabilities to different future locations, some more probable than others. This, to Einstein, didn't make sense. Another time he wrote:

> I find the idea quite intolerable that an electron exposed to radiation should choose of its own free will, not only its moment to jump off, but also its direction. In that case, I would rather be a cobbler, or even an employee in a gaming house, than a physicist.

There had to be something underneath, he believed, some hidden information that told the electron where to be, rather than pure randomly selected probability.

Einstein's toughest challenge to Neils Bohr and quantum theory came in the form of a thought experiment he dreamed up in 1935, usually referred to as the EPR experiment after the three people who wrote the paper, Einstein himself, Boris Podolsky, and Nathan

Rosen. The idea was simple, but the implications, which would result in the idea of quantum entanglement, were profound.

There were a number of ways to produce pairs of particles so that they were effectively twins. When you measured a particular property of one particle, for example its momentum or its spin, you also knew the same piece of information for the other particle; depending on what the property was, it would either have the same value or the diametrically opposite value. According to quantum theory, the particles didn't have a set value for this property until you measured it. It wasn't just that you didn't know the value until then, but it didn't exist. It's only at the point of taking the measurement that the value was fixed; until then it could have a whole range of values, each with a certain probability.

Now, here was the problem, as far as Einstein was concerned. Take two such particles and separate them to opposite sides of the universe. Now measure a property—say the spin—of one. Immediately, the other particle suddenly had to have the opposite value for that property. Instantly. Einstein had shown that nothing could travel faster than light, yet here the information of the property of one particle seemed to be communicating to the other vastly quicker than the speed of light, instantaneously. Einstein remarked to his friend and quantum physics supporter Max Born (not to be confused with Bohr or Bohm):

The whole thing is rather sloppily thought out, and for this I must respectfully clip your ear . . . whatever we regard as existing (real) should somehow be localized in time and space . . . [otherwise] one has to assume that the physically real in [position] B suffers a sudden change as a result of a measurement in [position] A. My instinct for physics bristles at this. However, if one abandons the assumption that what exists in different parts of space has its own, independent, real existence then I simply cannot see what it is that physics is meant to describe.

NONLOCAL REALITY

Quantum entanglement is the phenomenon that got David Bohm thinking about the way we look at quantum physics and wondering if there could be a different way to approach it. Initially there wasn't much need for this, as Einstein had just come up with this objection as a thought experiment. There was nothing practical that could be done with it. But since then, there have been remarkable developments in quantum entanglement which I cover in my book *The God Effect*.

Quantum entanglement enables us to produce unbreakable codes, build impossibly powerful computers, and even teleport particles from one place to another like a *Star Trek* transporter. It is a real, measurable effect. So either Einstein's doubts were wrong, or our interpretation of what was happening in quantum entanglement was mistaken. Bohm argued it was the interpretation that was at fault.

Even Einstein had allowed for this possibility, only to dismiss it. In that original EPR paper coming up with the concept of entanglement, he had not said that his thought experiment proved that quantum theory was wrong. He said that either quantum theory was wrong and there was hidden information that enabled the two particles at opposite ends of the universe to already know what value the property should have without communicating, or the whole idea of locality was rubbish. This second option Einstein dismisses, saying, "No reasonable definition of reality could be expected to permit this."

Locality. It's the kind of principle that is so obvious we usually assume it without even being aware of it. If we want to act on something that isn't directly connected to us, to give it a push, to pass a piece of information to it, or whatever, we need to get something from us to the object on which we wish to act. Often this "something" involves direct contact: I reach over and pick up my coffee cup to get it moving toward my mouth. But if we want to act on something at a distance without crossing the gap that separates us from that something, we need to send an intermediary from one place to another.

Imagine that you are throwing stones at a can that's perched on a fence. If you want to knock the can off, you can't just look at it and make it jump into the air by some sort of mystical influence; you have to throw a stone at it. Your hand pushes the stone and the stone travels through the air and hits the can; as long as your aim is good (and the can isn't wedged in place), the can falls off and you smile smugly.

Similarly, if I want to speak to someone across the other side of a room, my vocal cords vibrate, pushing against the nearest air molecules. These send a train of sound waves through the air, rippling molecules across the gap, until finally those vibrations get to the other person's ear, start her eardrum vibrating, and result in my voice being heard. In the first case, the stone was the intermediary, in the second the sound wave, but in both cases something traveled from A to B. This need for travel—travel that takes time—is what locality is all about. It says that you can't magically act on a remote object without that intervention.

Next time you are watching a magician at work, doing a trick where he manipulates an object at a distance, try to monitor your own reaction. As the magician's hand moves, so does the ball (or whatever the object he is controlling happens to be). Your mind rebels against the sight. You know that there has to be a trick. There has to be something linking the action of the hand and the movement of the object, whether directly, say with a very thin wire, or indirectly, perhaps by a hidden person moving the object while watching the magician's hand. Your brain is entirely convinced that action at a distance is not real.

Where there does seem to be action at a distance—a magnet attracting a piece of metal, or gravity pulling us toward the Earth, for instance—science explains it by saying that something is traveling between the two. Something invisible is carrying the force. However, these somethings don't break Einstein's relativity. We believe, for example, that gravity is transmitted at the speed of light. Quite different from the interaction in entanglement.

DISTANCE DOESN'T EXIST

When David Bohm looked at this problem he went at it a totally different way. Instead of worrying about the lack of locality he seized on this as a fundamental reality. If it is impossible for something to act instantly at a distance, he thought, why not do away with that distance entirely? Let's imagine that distance doesn't really exist. Suppose it's a concept that comes out of the way we interact with the universe, rather than a fundamental aspect of the universe's nature. Then with no distance between our two entangled particles there is no problem with the apparent communication. We could see them as two aspects of a complex whole, rather than the two truly individual entities, separated by space.

It's probable that Bohm was inspired to take the remarkable leap of reinterpreting the whole of reality by his experience with plasmas while still at Berkeley. As we have seen (page 106) plasma is the state of matter with a higher energy than a gas, in the same way as a gas is the state of matter with a higher energy than a liquid. In changing from liquid to gas, the atoms of a substance become more energetic, dancing around much more rapidly. In going from gas to plasma the atoms become more energetic still, blasting off their outer electrons, so a plasma is a gaseous soup of charged ions (atoms that have gained or lost electrons) and free electrons.

Bohm discovered that plasma can undergo an unusual effect now called Bohm diffusion that happened when the plasma was exposed to a magnetic field. He found that electrons seemed to lose their individuality under some circumstances, acting as if they were part of a connected whole.

Somewhat later, while he was at Princeton, his interest in plasma was extended to include the newly named plasmons. These are sometimes described as quasi-particles. It was discovered that the density of the electrons within a plasma could oscillate just like a wave, and it was possible to imagine this wave being broken up or quantized into quasi-particles, just as a light wave is actually quantized into photons. Plasmons are the quanta of this electron wave in the plasma.

When this happens, although the electrons still appear independent, they act as if they were all components of a single entity, resulting in the oscillation that forms the observed wave. For Bohm the way that the electrons acted together was fascinating, seeming to imply a sense of order that went beyond the obvious apparent collection of unconnected particles. From this Bohm would develop an alternative view of the physical world that required another field to be added to the standard collection.

PLAYING IN INVISIBLE FIELDS

Fields, a concept developed by Michael Faraday in the nineteenth century, are a way to look at how different objects influence each other at a distance. We often refer in physics to the four forces of nature: gravity, electromagnetism, the strong nuclear force, and the weak nuclear force. Each of these forces, it's thought, is transmitted from place to place by invisible intangible particles. The electromagnetic force, for example, is transmitted by photons.

However, we can also consider the force to be the result of a field, like a kind of invisible influence that spreads out from the object through space. Bohm suggested that there was another field, one that we weren't yet aware of, called the quantum potential, which unlike the force-based fields did not get weaker over distance. This quantum potential explained the way that particles could seemingly interact at a huge distance, because to the quantum potential the distance was meaningless.

You can see a similar kind of concept in the way the seventeenth-century philosopher René Descartes looked at light. Descartes wanted to explain how light got to the eye from a distant source such as a star. He envisaged an invisible "something" that filled empty space, the equivalent of a field, which he called the plenum. Light, he thought, was a "tendency to motion" in the plenum, resulting in pressure on the eyeball that generated the perceived light. Descartes' theory, although decidedly shaky, is often considered the starting point of the modern science of light, as it involves only the source and the medium through which light is transmitted, something that had never been stated explicitly before.

If Descartes had been right it would have meant that light had to move instantly from its source to the eyeball. It's as if there were a huge invisible pool cue stretching all the way from the star to the eye that views it. When the star pushes one end of the cue, instantly the other end pushes against the eye. The light doesn't take any time to arrive because, effectively, it is in all the places along the route at once.

In practice, Descartes thought the "pool cue," his plenum, consisted of a vast number of tiny, inflexible, invisible spheres. He imagined all of space filled with these minuscule balls. Pressure on one ball would be transmitted through millions of others before reaching its destination. The spheres would act as if there were a single object linking cause and effect. In Bohm's picture this crude collection of spheres becomes the quantum potential.

One of the more impressive things about Bohm's quantum potential is that it did away with Einstein's problem about locality. With a quantum potential viewpoint, there was no concept of locality. As did Descartes' plenum, the quantum potential provided a mechanism for objects to be linked with no concept of whether they were local to each other.

Bohm describes this quite poetically. When referring to plasmons he said, "through the action of the quantum potential, the whole system is undergoing a co-ordinated movement more like a ballet dance than like a crowd of unorganized people."

It was a simple physical model seen on a TV show that pushed Bohm into thinking about what might lie beneath his ideas of how the universe was linked by the quantum potential. The demonstration involved no high technology. It was just a glass jar with a big rotatable cylinder inside that only left a small gap between the jar and the cylinder. That gap was partly filled with the thick, viscous, transparent fluid glycerine. A small drop of ink was dropped onto the transparent substance; then the rest of the gap was filled up with more glycerine.

When the center cylinder was turned, the ink drop was spread out in the glycerine, becoming such a faint trace that it disappeared. But here's the bit that caught Bohm's attention. When the cylinder was carefully turned back, the ink reassembled itself into a single visible spot. Although not visible to the naked eye, the combination of the diffuse spread of ink and the device held within it the information needed to reassemble that ink spot.

UNIVERSE AS HOLOGRAM

Soon after seeing this, Bohm came across holograms for the first time, and here, he felt, was an even better picture of what he believed was happening in the universe. Take a look at the glass plate of the hologram itself and all you see is a random-seeming spatter of blotches and dots and fringes, a mess that has no obvious order to it. Yet illuminate it correctly and out pops the visible reality of the scene in all its three-dimensional glory. What, Bohm thought, if the universe were rather like that? A vast hologram comprising an uncountable set of tiny bits of information that only became the universe when taken together as a whole.

Here was a picture of the cosmos that wasn't a set of totally independent objects, but rather an incredibly complex whole whose structure was identified in a fashion that bore no resemblance to what we see when we look at the universe. You could almost say it represented the whole universe as a single, vast, ever-changing number.

Of course Bohm's idea of a holographic universe was far more than a traditional hologram. A hologram is just a single still picture of a particular point in time of a small proportion of space. Bohm's idea was something that encompassed the whole universe and that was constantly changing as particles interacted, something he saw as not being individual collisions but rather different manipulations in the structure he called the holomovement, to emphasize that it was much more than a static hologram.

In Bohm's universe there are no individual particles; everything is part of the same thing. This doesn't mean everything we see is meaningless and fake, just that the way it is made up is fluid rather than consisting of vast numbers of individual particles. Bohm used the example of structures that emerge in water, from waves to whirlpools. They certainly exist and can act on each other, and on other things, yet they aren't entities in their own right, separable from the water. They are all a part of the water "hologram."

AN INCOMPLETE PICTURE

Although Bohm's theory had a lot more detail than can be presented here, it isn't as complete as the mainstream ideas of particles and fields; nonetheless there are indications from experiment that suggest that it is worth considering further. However much Einstein didn't like it, all the work on quantum entanglement since his 1935 paper, both in taking measurements and applications such as quantum computers and quantum teleportation, show it exists. There really does seem to be some way for remote particles to communicate instantly, or to act as if they are part of the same thing, and Bohm's ideas have a remarkably good fit with this reality.

Bohm's idea of this universal hologram of interaction has never been satisfactorily taken forward. Most physicists still find it difficult to deal with, and consider it more an interpretation than something that can be used in a practical theory. The standard approach to quantum theory still makes the numbers work well and made it possible to develop electronics, a technology totally built on quantum physics. Yet it can't be denied that there is a certain appeal to Bohm's theory, which puts forward a view of a universe in which "before the Big Bang" becomes just a different aspect of the holographic whole. It allows for a continuum of existence in which the Big Bang is just a passing milestone, rather than a true beginning.

A few scientists have taken the idea even further. Leading physicist Roger Penrose believes that consciousness is also a holographic concept. Our brains don't store memory in a simple bit-by-bit way as does computer memory, nor can we easily explain consciousness in such a reductionist way. Individual cells in the brain don't tell us anything about consciousness: it has to be treated as a whole, and it's easy to see it would be possible to apply a similar holographic model for the relationship between the brain and consciousness as Bohm was suggesting for the relationship between the "real" holographic universe and what we observe.

This was the picture of the way the conscious brain operated that neurophysiologist Karl Pribram from Stanford University developed independently of Bohm's work, and the two fit together well. However, others have taken Bohm's ideas to provide an explanation for paranormal events that have yet to be proved to exist, from remote viewing to miracle working, or to give weight to Eastern religious faiths that take the view that everything is part of the same whole. Perhaps the worst thing about Bohm's theories is the way that they can be used to pull in all sorts of confused thinking masquerading as science.

This appears to be the case with the late-twentieth-century philosophical doctrine of postmodernism. Postmodernism has always disliked science, because this almost meaningless philosophy doesn't accept the value of reductionism. Science usually works by breaking something down to its component parts and seeing how it functions. Those who don't like reductionism say this misses the whole point of the way everything is part of a whole, and separating off parts loses the point.

However, the problem with such anti-reductionism is that it doesn't really reflect what we see in the world. You can say that perceptions are more important than "reality" and natural laws are just manmade impositions as much as you like. But the fact is, if you walk off the top of a twenty-story building, your perceptions that you can levitate won't help you in the slightest. Gravity will win. It's a mistake to class Bohm's ideas with this kind of thinking, because although they depend on a holistic picture of the universe, they encompass reductionist results, rather than denying them.

Yes, Bohm said everything was interlinked and the whole was what mattered, but he didn't deny that the outcome of this whole was something that has an impact at the level of what we consider to be individual particles. Otherwise it's a bit like saying, "The sea is a big friendly thing; waves are just insignificant ripples relative to the whole," as a tsunami hurtles toward your beach. Those insignificant ripples can still tear solid physical objects apart.

THE FUTURE OF THE PAST

And so in our trip through Bohm's amazing holographic universe we see one picture of reality where the question, What happened before the Big Bang? requires us to take a different view of the whole universe from the one we normally perceive, describing our conventional observations as an illusion, a limited view onto a much grander and entirely interlinked whole.

As scientists fathom deeper the possibilities for the cosmos, whether with universe-spanning telescopes, the power of vast machines such as the Large Hadron Collider, or just the age-old tools of the theoretical physicist—a pad, a pencil, and an imaginative brain—we can never be sure that we will have a certain answer to the key question posed by this book.

We know that there are some questions in math that it is impossible to prove either way. This strange conclusion was reached by the eccentric mathematician Kurt Gödel. Although Gödel proved this result mathematically, we can get a similar feel from the logical paradox made famous by philosopher Bertrand Russell. The village barber cuts everyone's hair except people who cut their own hair. Does the barber cut his own hair? Because the answer has to be yes and no, we are forced to concede that the problem is impossible to answer. There are some formal mathematical propositions that it can be shown are no more possible to solve.

Similarly, the computer pioneer Alan Turing, one of the team who cracked the German Enigma code during the Second World War, proved that there were some problems that could not be solved by computer, no matter how fast it was.

It may be that there are some scientific questions that fall into the same category. Questions that in principle have answers, but in practice are never satisfactorily answered. Some are science-fiction classics. Can we build a time machine? Can we travel faster than light? Physicists have come up with theoretical solutions to these problems (they're linked) but they may well never be made real. Similarly we may never have a definitive answer to the question, What came before the Big Bang?

Personally, I find myself in a real quandary. I very much like Turok and Steinhardt's bouncing brane theory; it has a feeling of elegance that the much-patched and fudged Big Bang plus inflation theory doesn't. Yet bouncing branes are dependent on M theory with the baggage and worries about the validity of string theory that it brings with it. It seems whichever way you turn, there is no easy answer when it comes to the earliest moments of the universe.

In the future we may discard all our current theories and come up with something new and different, although it would have to incorporate the increasingly rich data that are used to support current theories. Or we may continue to refine existing ideas as more information comes in, never able to break through those final barriers of certainty. Even so, the exercise is a fascinating one, and well worth undertaking.

We may not have definitive answers, but the different possibilities remain a delight that will intrigue anyone who looks out at the night sky with a sense of awe, wondering where it all came from and where it all began.

NOTES

2. ENTER THE CREATOR

8. Arthur C. Clarke's comment on technology and magic first appeared in Arthur C. Clarke, *Profiles of the Future: An Enquiry into the Limits of the Possible*, London: Gollancz, 1982.

16. Roger Bacon's views on the mechanism of the stars and Moon are described in Brian Clegg, *The First Scientist*, London: Constable and Robinson, 2003.

18. Information on the greenhouse effect is from Brian Clegg, *Ecologic*, London: Transworld, 2009.

18. Fred Adams' suggestion that many possible universes could have energy sources to support life is described in Michael Brooks, "In the Multiverse Stars Burn Black," *New Scientist*, August 2, 2008.

3. WHAT AND HOW BIG?

20. Definition and etymology of "universe" is taken from *The Oxford English Dictionary*, Oxford: Oxford University Press, 1989.

22. Quotes from Archimedes' *The Sand Reckoner* are from Archimedes, *The Works of Archimedes* (Ed: T. L. Heath), New York: Dover, 2003.

32. Details of Richard Bentley's life are from James Henry Monk, *The Life of Richard Bentley*, London: J. G. & F. Rivington, 1833 (Available on Google Books http://books.google.co.uk/books?id=0UoJAAAAQAAJ).

35. The study showing 70 percent of textbooks explained Olbers' paradox incorrectly in 1987 is quoted in Michio Kaku, *Parallel Worlds*, London: Penguin, 2006.

36. Poe's solution to Olbers' paradox is found in Edgar Allan Poe, *Eureka: A Prose Poem*, available widely online, including http://xroads.virginia.edu/~hyper/poe/eureka.html.

38. Herschel's biographical details are from Brian Clegg, *Light Years*, London: Macmillan Science, 2007.

41. Herschel's measurement of the universe's size in siriometers is described in Simon Singh, *Big Bang*, London: Harper Perennial, 2005.

41. Distance to Sirius is taken from Patrick Moore, *The Amateur Astronomer*, London: Lutterworth, 1957.

43. Bessel's early use of parallax to measure stellar distance is described in Simon Singh, *Big Bang*, London: Harper Perennial, 2005.

45. Information on Henrietta Leavitt's work is from George Johnson, *Miss Leavitt's Stars*, New York: W. W. Norton, 2006.

47. Among others, Simon Singh asserts incorrectly that "Herschel indicated that everything was in our pancake-shaped Milky Way" in Simon Singh, *Big Bang*, London: Harper Perennial, 2005.

48. The development of Herschel's theories of the nature of nebulae is detailed in C. A. Lubbock, *The Herschel Chronicle*, Cambridge: Cambridge University Press, 1933.

48. Ernest Rutherford's comment that all science is either physics or stamp collecting is quoted from J. B. Birks, *Rutherford at Manchester* (1962) in *The Oxford Dictionary of Scientific Quotations*, Oxford: Oxford University Press, 2005.

49. Assertion that no nebulae compare with the Milky Way is from Agnes M. Clerke, *The System of the Stars*, London: Adam & Charles Black, 1905.

50. Information on the development of telescopes between Herschel and Hubble is from Patrick Moore, *Eyes on the Universe: The Story of the Telescope*, London: Springer-Verlag, 1997.

58. Hubble's description of the universe as 6 billion light-years in diameter is described in *The Times* (London), April 26, 1934.

62. Boudewijn Roukema's demonstration that the universe could be a dodecahedron is described in Boudewijn F. Roukema et al., "The optimal phase of the generalised Poincaré dodecahedral space hypothesis implied by the spatial cross-correlation function of the WMAP sky maps," Arxiv.org, 0801.0006.

4. HOW OLD?

66. Herschel's comment about observing stars of which the light must have taken 2 million years is from C. A. Lubbock, *The Herschel Chronicle*, Cambridge: Cambridge University Press, 1933.

67. Information on the idea of the tipping point is from Malcolm Gladwell, *The Tipping Point*, London: Abacus, 2002.

67. Information on chaos theory is from James Gleick, *Chaos*, London: Penguin, 1988.

69. Arthur Eddington's British Association lecture on the expanding universe is covered in *The Times* (London), September 13, 1933.

77. Rutherford's pioneering work in radio-dating is described in Hamish Campbell, *Discovering the Age of the Earth*, part of *The Elegant Universe of Albert Einstein*, Wellington: Awa Press, 2006.

5. A BANG OR A WHIMPER?

84. Lemaître's use of cosmic rays to help corroborate his theory is described in Simon Singh, *Big Bang*, London: Harper Perennial, 2005.

90. Information on Huggins' use of spectroscopy to measure relative velocity in space is from Simon Singh, *Big Bang*, London: Harper Perennial, 2005.

91. Hubble and Humason's research on the movement of galaxies is described in Simon Singh, *Big Bang*, London: Harper Perennial, 2005.

93. Details of Isaac Newton's comments about not framing a hypothesis are from Brian Clegg, *The God Effect*, New York: St. Martin's Press, 2006.

95. Oliver Lodge's comment about Eddington understanding relativity is from *The Times* (London), December 13, 1919.

96. Eddington's count of the number of protons in the universe from his Tarner Lectures is quoted in *The Oxford Dictionary of Scientific Quotations*, Oxford: Oxford University Press, 2005.

112. Ernest Sternglass' theory that the universe originated in a supermassive electron/positron pair is described in Ernest J. Sternglass, *Before the Big Bang*, New York: Four Walls Eight Windows, 1997.

6. KEEPING THINGS STEADY

117. Biographical details of Fred Hoyle are from Jane Gregory, *Fred Hoyle's Universe*, Oxford: Oxford University Press, 2005.

118. Hoyle's attack on the Nobel Prize award for the discovery of the pulsar is covered in *The Times* (London), March 22, 1975, and April 8, 1975.

119. Stephen Hawking's comment that many people dislike the Big Bang because it smacks of divine intervention is from Stephen Hawking, *A Brief History of Time* (Twentieth Anniversary Commemorative Edition), London: Transworld, 2008.

121. The suggestion that those behind the Steady State theory used the movie *Dead of Night* only as a metaphor, not as an inspiration for their theory, is in Michio Kaku, *Parallel Worlds*, London: Penguin, 2006.

124. Hoyle's assertion that it doesn't matter how strange an idea is, is from Fred Hoyle, *The Nature of the Universe*, London: Penguin, 1963.

124. Feynman's assertion that common sense isn't a good test of a scientific theory is from Richard Feynman, *QED: The Strange Theory of Light and Matter*, London: Penguin, 1990.

129. B. Y. Mills' damning comments on the inability to use the Cambridge results to make a decision on cosmological matters are in B. Y. Mills and O. B. Slee, *Australian Journal of Physics*, 10, 162, 1957.

129. Fred Hoyle's feelings when Martin Ryle sprang data on him at a press conference are from Fred Hoyle, Geoffrey Burbidge, and Jayant Narlikar, *A Different Approach to Cosmology*, Cambridge: Cambridge University Press, 2000.

132. For quite technical details of Hoyle's quasi–Steady State model, see Fred Hoyle, Geoffrey Burbidge, and Jayant Narlikar, *A Different Approach to Cosmology*, Cambridge: Cambridge University Press, 2000.

7. INFLATING THE TRUTH

141. Details of Robert Dicke's realization that there should be background radiation from the remnants of the Big Bang is described in Marcus Chown, *Afterglow of Creation*, London: Arrow Books, 1993.

143. The revelation that the effects of background radiation had been spotted indirectly before Gamow predicted its existence is from Marcus Chown, *Afterglow of Creation*, London: Arrow Books, 1993.

147. The variation in cosmic background radiation due to the relative movement of the Earth is described in Paul Davies, *The Goldilocks Enigma*, London: Penguin, 2007.

150. Simon Singh's assertion that detecting the cosmic microwave background radiation would "prove that the Big Bang really happened" is in Simon Singh, *Big Bang*, London: Harper Perennial, 2005.

151. Sakharov's theory that all but one in a billion of the matter particles after the Big Bang were annihilated by anti-matter is described in Martin Rees, *Our Cosmic Habitat*, London: Orion, 2002.

151. The idea that the anti-matter from the early days of the universe could still exist in pockets comes from Neil deGrasse Tyson and Donald Goldsmith, *Origins*, New York: W. W. Norton, 2004.

153. The latest paper giving support to inflation by using the "sum across possible universe" approach is J. B. Hartle, S. W. Hawking, and T. Hertog, "No-boundary measure of the universe," *Physical Review Letters* 100, 201301, 2008.

154. Benjamin Wandelt's paper disputing evidence for inflation in the universe is A. P. S. Yadev and B. D. Wandelt, "Evidence of primordial non-Gaussianity (f_{NL}) in the Wilkinson microwave anisotropy probe 3-year data at 2.8σ," *Physical Review Letters* 100, 181301, 2008.

155. Issues with inflation are described in Michael Brooks, "Inflation Deflated," *New Scientist*, June 7, 2008.

156. Details of the incorrect quantities of lithium predicted by the Big Bang theory are from Matthew Chalmers, "Crucible of Creation," *New Scientist*, July 5, 2008.

159. Information on Young's work on the wave nature of light and the dismissal of the ether is from Brian Clegg, *Light Years*, London: Macmillan Science, 2007.

161. The suggestion that dark matter is a different kind of assumption from the ether is made in Neil deGrasse Tyson and Donald Goldsmith, *Origins*, New York: W. W. Norton, 2004.

167. Information on the cosmological constant (Einstein's greatest mistake) is from Amir Aczel, *God's Equation: Einstein, Relativity and the Expanding Universe*, London: Piatkus, 2000.

8. LET THERE BE TIME

177. Information on St. Augustine of Hippo and quotes from his *Confessions* are taken from St. Augustine (translated, edited, and introduced by Henry Chadwick), *Confessions*, Oxford: Oxford University Press, 1998.

9. GROUNDHOG UNIVERSE

188. Robert Dicke's development of the concept of a cyclic universe with both the Big Bang and Big Crunch is described in Marcus Chown, *Afterglow of Creation*, London: Arrow Books, 1993.

190. Richard Tolman's demonstration that a conventional cyclic universe would have a finite earlier life is described in Paul J. Steinhardt and Neil Turok, *Endless Universe: Beyond the Big Bang*, London: Weidenfeld and Nicolson, 2007.

200. Michio Kaku's concern that so much time might have been wasted on string theory is from Michio Kaku, "Unifying the Universe," *New Scientist*, April 16, 2005.

200. Lee Smolin's comment about career suicide is from Lee Smolin, *The Trouble with Physics*, London: Penguin, 2007.

201. Lee Smolin and Richard Feynman's concerns about string theory are from Lee Smolin, *The Trouble with Physics*, London: Penguin, 2007.

201. For more on the reasons that string theory could not only be wrong, but not even science, see Lee Smolin, *The Trouble with Physics*, London: Penguin, 2008.

202. The comment that string and M theorists could be "away in Never-Never Land" is from Paul Davies, *The Goldilocks Enigma*, London: Penguin, 2007.

203. For more on string theory and potential experimental verification, see Brian Greene, *The Elegant Universe*, London: Vintage, 2000.

205. Michell's early ideas of black holes and Wheeler's naming of them is described in Stephen Hawking, *A Brief History of Time* (Twentieth Anniversary Commemorative Edition), London: Transworld, 2008.

209. Turok and Steinhardt's concept of a cyclic universe driven by bouncing branes is described in Paul J. Steinhardt and Neil Turok, *Endless Universe: Beyond the Big Bang*, London: Weidenfeld and Nicolson, 2007.

210 Peter Woit calls string theory and M theory "not even wrong" in Peter Woit, *Not Even Wrong: The Failure of String Theory and the Continuing Challenge to Unify the Laws of Physics*, London: Jonathan Cape, 2006.

218. Cristiano Germani's idea of the universe being a brane traveling back out of a throat in a Calabi-Yau manifold is described in Zeeya Merali, "Bye-bye Big Bang, Adios Inflation," *New Scientist*, September 8, 2007.

220. Bouncing models of the universe, including Martin Bojowald's model with "cosmic forgetfulness" are described in Charles Q. Choi, "New Beginnings," *Scientific American*, October 2007.

220. Martin Bojowald's cosmic forgetfulness ideas are described in Martin Bojowald, "What Happened before the Big Bang," *Nature Physics*, 3, 525, 2007.

221. Parampreet Singh and Alejandro Corichi's modeling of the pre–Big Bang universe using loop quantum gravity is described in Alejandro Corichi and Parampreet Singh, "Quantum Bounce and Cosmic Recall," *Physical Review Letters*, 100, 161302, 2008.

10. LIVING IN A BUBBLE

227. The idea that inflation could have gone on forever without any beginning is suggested by Andrei Linde in S. W. Hawking and W. Israel (eds.), *300 Years of Gravitation*, Cambridge: Cambridge University Press, 1987.

232. Lee Smolin's evolution of universes idea is described in Michio Kaku, *Parallel Worlds*, London: Penguin, 2006.

233. The black hole mechanism of Smolin's evolutionary universes is described in Martin Rees, *Our Cosmic Habitat*, London: Orion, 2002.

234. Paul Steinhardt's refusal to even contemplate the multiverse concept is quoted in Paul Davies, *The Goldilocks Enigma*, London: Penguin, 2007.

236. The Arkleseizure quote is taken from Douglas Adams, *The Hitchhiker's Guide to the Galaxy—The Original Radio Scripts*, London: Pan, 1985.

236. The short story "Neutron Tide" appears in Arthur C. Clarke, *The Collected Stories*, London: Gollancz, 2001.

242. Anatoly Svidzinsky's suggestion that clouds of dark matter provide the mass at the center of galaxies, not supermassive black holes, is from Zeeya Merali, "Bubble Ousts Black Hole at Center of the Galaxy," *New Scientist,* July 27, 2006.

245. CERN's dismissal of the dangers of the Large Hadron Collider is from its Web site at http://press.web.cern.ch/public/en/LHC/Safety-en.html.

245. Michio Kaku's speculation about the potential of future civilizations to travel between universes is from Michio Kaku, *Parallel Worlds*, London: Penguin, 2006.

11. WELCOME TO THE MATRIX

249. Seth Lloyd's statement about the "it from bit" concept is from Seth Lloyd, *Programming the Universe*, London: Vintage Books, 2007.

254. For more on quantum computers, see Brian Clegg, *The God Effect*, New York: St. Martin's Press, 2006.

12. SNAPSHOT UNIVERSE

265. Information on holograms is taken from Brian Clegg, *Light Years*, London: Macmillan Science, 2007.

269. Information on stereographic photography is taken from Brian Clegg, *The Man Who Stopped Time*, Washington: Joseph Henry Press, 2007.

272. For a highly technical exploration of the black hole version of the holographic universe, see Leonard Susskind and James Lindesay, *An Introduction to Black Holes, Information and the String Theory Revolution—The Holographic Universe*, Hackensack: World Scientific, 2005.

273. Arthur Eddington's remarks on the second law of thermodynamics are from his 1927 Gifford Lecture, written up in *The Nature of the Physical World* (1928) and reported in *The Oxford Dictionary of Scientific Quotations*, Oxford: Oxford University Press, 2005.

276. David Bohm's ideas on the holographic universe are described in Michael Talbot, *The Holographic Universe*, London: HarperCollins, 1991.

278. Information on Einstein's problems with quantum theory, the EPR experiment, and quantum entanglement is from Brian Clegg, *The God Effect*, New York: St. Martin's Press, 2006.

278. The "EPR" paper is A. Einstein, B. Podolsky, and N. Rosen, "Can Quantum-Mechanical Description of Physical Reality Be Considered Complete?", *Physical Review*, 47, May 15, 1935.

285. Bohm's comments about the coordination of electrons in a plasma by the quantum potential are from David Bohm, *Hidden Variables and the Implicate Order* in Basil J. Hiley and F. David Peat, *Quantum Implications: Essays in Honour of David Bohm*, London: Routledge, 1991.

INDEX

SCIENCE. SO YOU CAN UNDERSTAND IT.

THE GOD EFFECT
QUANTUM ENTANGLEMENT, SCIENCE'S STRANGEST PHENOMENON

"A clear, nontechnical guide to entanglement and its fascinating implications." —PHYSICS WORLD

TIME TRAVEL, TELEPORTATION, AND THE ULTIMATE COMPUTER

BRIAN CLEGG

ARMAGEDDON SCIENCE
THE SCIENCE OF MASS DESTRUCTION
BRIAN CLEGG

A look at quantum entanglement—the physical phenomenon that allows two particles to project instantaneous communication to one other.

An authoritative look at real mad science at work today, recklessly putting life on Earth at risk for the pursuit of knowledge and personal gain.

www.stmartins.com
www.brianclegg.net

Available wherever books are sold